Regulating Toxic Substances

ENVIRONMENTAL ETHICS AND SCIENCE POLICY SERIES

Kristin Shrader-Frechette, General Editor

Acceptable Evidence:
Science and Values in Risk Management
Edited by Deborah Mayo and Rachelle D. Hollander

Experts in Uncertainty:
Opinion and Subjective Probability in Science
Roger M. Cooke

Regulating Toxic Substances:
A Philosophy of Science and the Law
Carl F. Cranor

Regulating
Toxic Substances

A Philosophy of Science
and the Law

Carl F. Cranor

New York   Oxford
OXFORD UNIVERSITY PRESS
1993

Oxford University Press

Oxford   New York   Toronto
Delhi   Bombay   Calcutta   Madras   Karachi
Kuala Lumpur   Singapore   Hong Kong   Tokyo
Nairobi   Dar es Salaam   Cape Town
Melbourne   Auckland

and associated companies in
Berlin   Ibadan

Copyright © 1993 by Carl F. Cranor

Published by Oxford University Press, Inc.
200 Madison Avenue, New York, NY 10016

Oxford is a registered trademark of Oxford University Press

Library of Congress Cataloging-In-Publication Data
Cranor, Carl F.
Regulating toxic substances : a philosophy of science
and the law / Carl F. Cranor.
p. cm.—(Environmental ethics and science policy series)
Includes bibliographical references and index.
ISBN 0-19-507436-X
1. Hazardous substances—Law and legislation—United States.
2. Chemicals—Law and legislation—United States.
3. Industrial safety—Law and legislation—United States.
I. Title.
II. Series: Environmental ethics and science policy.
KF3958.C73   1993   344.73′04622—dc20
[347.30446.22]
91–47046

10 9 8 7 6 5 4 3 2 1
Printed in the United States of America
on acid-free paper

For Crystal and Christopher,
who waited,
and
for Valerie and Nathan

Foreword

The use and misuse of science in the policy process is increasing as many of the problems pressing upon government institutions are being driven by technological advances. Legislative and regulatory assumptions made years ago are being constantly challenged as technological advances allow more sensitive measurement of toxic substances, present new and novel risks, or reverse commonly held scientific beliefs about the causal relationships between hazards and health effects.

Policy makers and judges are forced to deliberate about issues which involve scientific complexities beyond the understanding of most jurists or elected officials. Expert witness fees are becoming a predictable supplement to the incomes of university researchers as epidemiologists, geneticists, toxicologists, and other scientists are called upon in court proceedings. Legislative debates are shot through with technical information as science is bent one direction and another to support positions being taken. And as the use of science in policy arenas increases, there is corresponding pressure to enact laws which keep pace with the science upon which the policy is based.

But where does this process lead us? When one reviews the simple elegance of older statutes, one is impressed with the clarity and directness of the law. Modern legislation, especially regulatory legislation, is cluttered with a level of technical detail formerly left to the executive branch or the courts, or to the scientist working in his or her laboratory. Lost in this detail is the central thrust of the law and an appeal for simple solutions which reflect the mores of society.

What is evolving from this process is a basic conflict between the quantitative rigor of science and scientific information and the qualitative and flexible nature of the policy process. Even the judicial process, which is based upon precedent and a predictable extrapolation of prior decisions, and thereby claims a greater rationality than legislative process, is political and responds to changing public opinion and values. This conflict forces us to reflect upon how we should best accommodate and respond to increasingly technical information consistent with our legal and social policy goals.

At the same time, modern legislators and regulatory managers long for a scientific process which will produce irrefutable results, eliminating the grey areas of policy decisions. While publicly deriding so-called "bright line" legislative or policy standards, such as the Delaney Clause in food law, legislators and bureaucrats alike find some refuge in the inflexibility of these standards. With absolute standards, difficult risk-benefit discussions and painful policy debates about "acceptable" cancer rates are eliminated. Absent the political safety of these absolutes, the debate centers on the level of certainty represented by the scientific arguments being presented to support each side of an issue.

But scientific certainty in a regulatory process is at present an illusory goal, the dogged pursuit of which can actually impede the use of a solid scientific approach and frustrate legal goals. If the pursuit of certainty leads to the need for solid proof before regulatory action is initiated, then the regulatory process will bog down. If there is a need to prove with absolute certainty the link between exposure to a substance and a human health effect before any action is taken, then commencement of needed risk management and preventive health measures may be delayed.

Rigid systems are being developed in response to rigid and specific statutory mandates or in response to fears of scientific challenges from those adversely affected. For some regulatory programs, this is leading to paralysis.

In the pesticide regulatory program at the Environmental Protection Agency (EPA), a rigid regulatory approach has prevented EPA from taking a number of innovative approaches to regulation. Aided by Congressional inability to modernize either of the statutes affecting food safety regulation, EPA has labored chemical-by-chemical over the last two decades and still faces major backlogs in reviewing currently registered pesticides. This situation is changing, as EPA begins to move in the direction of pollution prevention and a "safer" pesticide policy, but major obstacles remain.

However, this situation need not be the rule. California's Proposition 65, a fairly rigid state initiative dealing with carcinogens, has been implemented in a fair manner which has quieted the fears of most of its critics. Negotiated rulemaking at the EPA has provided examples of how a political consensus can help overcome a scientific difference of opinion. But there need to be more examples of innovative regulatory behavior.

Before this can occur, we need to look at the use of science in the policy and regulatory arenas in a new light. We need to decide how much scientific certainty we demand before taking regulatory action and how rapidly we assess and regulate potential health risks, issues which are philosophical and policy questions of considerable importance to our health and well-being. We need to challenge some of our assumptions about the processes we use, processes which, for the most part, have been developed incrementally. This is the contribution that this book makes.

The ideas raised by Dr. Cranor provoke thought and bring a philosopher's view, a normative view, to the problem. The basic challenge to the current process raised in this book deserves some attention and debate: whether we ought to require the same scientific rigor in the legal process which we would apply in the laboratory. We need to decide whether we can afford our present, time-consuming procedures for evaluating potentially toxic substances when other procedures are available which might better serve our policy goals.

Since much of our current policy and regulatory paralysis results from scientific stalemates or from delays pending the result of the "definitive" test results, perhaps we should be looking at some alternatives, such as those suggested here. Especially for our pollution prevention and preventive health effects efforts, we should be developing a quicker response threshold.

This debate on the need for scientific certainty in governmental actions will continue for years, in Congress, in the courts, and in the regulatory agencies. We should all take some time out from that debate to wonder if there aren't some other, simplified and expedited approaches which would better meet our social goals than the ones we are using. We should take time to reflect on the issues raised in this book.

The Honorable George E. Brown, Jr.
Chairman, Committee on Science, Space, and Technology
U.S. House of Representatives

Preface

My interest in environmental issues began early, in the relatively pristine environment of Colorado's Rocky Mountains, where I was fortunate to be born and raised. Initially I took a clean and nonthreatening environment for granted, because even though that particular natural world contained threats of its own—freezing winters, avalanches, wild animals—the air and water typically were not harmful to health. In fact, they were among the cleanest to be found. When I moved to Los Angeles, I began to appreciate the health effects of toxic substances. The risks posed by air, water, workplace, and even food were as dangerous as the obvious and palpable threats of the Rockies. My philosophical interests in environmental issues, however, did not begin until much later.

Theoretical normative ethics and legal philosophy were the focus of my early work. I sought to understand and articulate principles that protect the individual from the adverse effects of policies, procedures, and institutions. These earlier concerns motivate some of the views in this book. Upon exploring environmental health issues, I found that much of the normative discussion, dominated by economists and others, emphasizes aggregate human welfare and tends to lose sight of the effects of policies on the lives of individual persons. Yet individuals can suffer substantial harm from toxic substances that are wrongly identified as safe or that are underregulated. Such effects on people raise questions about justice.

In the late 1970s, while my interests in normative ethics continued, I wanted to bring together my quasi-avocational concerns for environmental health protection and my interests in legal and moral philosophy. The subjects that interested me were the ways in which human beings can harm one another with their molecules, that is, with toxic substances. These issues are particularly difficult, for harms from toxic substances are unlike other injuries that we inflict on one another with weapons or cars or explosions where the harms are typically immediate and obvious. The damage from invisible toxins can, of course, be just as great or greater to an individual, but it may remain hidden for many years. And whether there is harm at all is frequently an open question. There is much uncertainty about whether substances are harmful, the severity of harms they cause, the mechanisms by which injuries are produced, and how best to control exposure to such substances, especially when they are beneficial to us.

The first step in pursuing this new research was to learn the appropriate environmental, tort, and administrative law at Yale Law School in 1980–81 as a Master of Studies in Law Fellow supported by an American Council of Learned Studies Fellowship. During the 1980 term the Supreme Court announced its *Benzene* case decision, *Industrial Union Department, AFL-CIO, v. American Petroleum Institute*, 448 U.S. 607 (1980). I read this decision, eventually publishing some papers on it.

The case revealed that court misunderstandings about some of the scientific tools available for regulating toxic substances could substantially, but perhaps inadvertently, bias decisions against one side or another in a legal controversy.

The next step was to acquire a better understanding of the details of the regulatory law and of the science that leads to the regulation of toxic substances. A Congressional Fellowship from the American Philosophical Association during 1985–86 provided the opportunity. For one-fourth of that year I had the opportunity to observe congressional lawmaking as a legislative aide to Congressman George E. Brown, Jr., a senior member (and now chairman) of the House Committee on Science, Space and Technology. The remainder of the time, at the U.S. Congress's Office of Technology Assessment, I worked on *Identifying and Regulating Carcinogens,* a comprehensive review of federal agency efforts to identify and regulate carcinogens as required by a number of laws passed by Congress in the 1970s. Both experiences provided an opportunity to observe major aspects of environmental health policy in the making.

It became apparent that seemingly factual and "scientific" decisions concealed in the bowels of administrative agencies' risk assessment policies and decisions could have substantial impacts on the health of our citizenry and the wealth of the country. Demanding too much scientific evidence could leave potentially toxic substances unidentified or underregulated; accepting too little could potentially affect the overall wealth and economic well-being of the country.

Upon returning to the university, I sought to understand better the scientific and normative underpinnings of risk assessment procedures. The University of California's Toxic Substances Research and Teaching Program funded much of this research. An interdisciplinary research group comprised of an environmental engineer, a resource economist, a toxicologist/biochemist, a statistician, and a philosopher evaluated scientific and policy aspects of carcinogen risk assessment. Our research confirmed concerns about the effects of carcinogen risk assessment practices on legal and normative questions. Recommendations by many in the scientific, regulated, and regulatory community to adopt more detailed risk assessments appeared to exacerbate existing problems. Alternative procedures seemed to be needed. Continued support from the National Science Foundation permitted me to address some of the legal and institutional issues.

A 1989 appointment to California's Proposition 65 Science Advisory Panel provided me with firsthand knowledge of the regulatory and risk assessment process. That panel, created by a voter-passed initiative, the Safe Drinking Water Act of 1986, is required to identify and list for various legal purposes carcinogenic and reproductive toxins. This experience, as well as the opportunity to work with members of the California Department of Health Services (now the California Environmental Protection Agency), further highlighted the importance of examining critically the use of scientific evidence and its effects in our legal institutions.

In the spring of 1990 both the Proposition 65 Science Advisory Panel and the Science Advisory Board of the California Air Resources Board began to explore the idea of expedited risk assessment procedures for the reasons given in Chapter 4 of this book. In April 1991 the Proposition 65 Science Advisory Panel unanimously recommended the adoption of such procedures; they are now being considered for regulation.

What began as a concern about the use and role of scientific evidence in the law and its importance to our lives has now been clarified and developed, informed both by research and by practical exposure to some of the issues. The scientific tools and evidentiary procedures used for legal purposes can inadvertently affect the regulation of toxic substances and their impact on human beings in ways we might not anticipate. An institution's approach to using scientific evidence raises substantial philosophical questions important to the outcome of public policy. Yet for the most part these issues have been decided with virtually no philosophical input and frequently with no public scrutiny. Although this is still largely the case, I hope this book brings some of the issues to scholarly and public debate.

Riverside, Calif. C. F. C.
December 1991

Acknowledgments

I owe debts of gratitude to numerous people and agencies for the work produced here. Research leading to this book has been supported by the American Council of Learned Societies, Yale Law School, the American Philosophic Association's Congressional Fellowship Program, The University of California's Toxic Substances Research and Teaching Program, the National Science Foundation (Grant DIR-8912780), and the University of California, Riverside, Intramural and Intercampus Research Grant Program. This funding supported numerous graduate students and departmental staff, who made the work much easier than it otherwise would have been with my other administrative and public duties over the past several years. In particular Kenneth Dickey worked as a research assistant on the UCR Carcinogen Risk Assessment Project from its inception in 1987 to 1990; he was invaluable in that capacity. More recently Sara Austin, Steven Pfeffer, Kerrin McMahan, and Andrew Light provided valuable support, which is much appreciated.

Several institutions with which I had the opportunity to work have improved my understanding of the issues: the U.S. Congress, its Office of Technology Assessment, and the California Department of Health Services (now known as the California Environmental Protection Agency). I am grateful for these experiences.

I am grateful to many from whom I have learned about this subject. The manuscript benefited from advice and criticism from members of the UCLA Law and Philosophy Discussion Group, which includes Steve Munzer, David Dolinko, Peter Aranella, Jean Hampton, Craig Ihara, Kurt Nutting, David Copp, John Fischer, and Mark Ravizza, and from discussions with my colleagues Larry Wright and Alex Rosenberg. I have also learned much from Craig Byus, Andrew Chang, D. V. Gokhale, and Lars Olson, co-principal investigators on the UCR Carcinogen Risk Assessment Project. Comments from Oxford University Press referees have improved the book further.

Portions of Chapter 1 are adapted from previous work in "Epidemiology and Procedural Protections for Workplace Health in the Aftermath of the *Benzene* Case," *Industrial Relations Law Journal* 5(3) (1984):372–401; "Some Moral Issues in Risk Assessment," *Ethics* 101 (October 1990):123–143, The University of Chicago Press; "Scientific Conventions, Ethics, and Legal Institutions," *Risk* I (1990):155–184; and "Some Public Policy Problems with the Science of Carcinogen Risk Assessment," *PSA 1988,* Vol. 2, pp. 467–88. Sections of Chapter 2 draw from "Scientific and Legal Standards of Statistical Evidence in Toxic Tort and Discrimination Suits," *Law and Philosophy* 9:115–156, 1990 (with Kurt Nutting). Parts of Chapter 3 utilize and adapt material from "Joint Causation, Torts and Regulatory

Law in Workplace Health Protections,'' *The International Journal of Applied Philosophy*, Vol 2, No. 4 (Fall 1985), pp. 59–84. Finally, some sections of Chapter 5 are modified versions of ''Justice in Workplace Health Protections,'' *Logos* (1985), pp. 131–147.

Finally, I want to thank Gretchen Huth and Deidra Price for their assistance on the project over the years and Vicki Rickard, Cindi Smith, and Audrey Lockwood for entering many pages of this book into the word processor. Parts of this manuscript appeared in early versions in *Ethics, Law and Philosophy, The Industrial Relations Law Journal, PSA 1988, The Yale Law Journal, The International Journal of Applied Philosophy, Logos,* and *Risk.*

Riverside, Calif.   C. F. C.
December 1991

Contents

Regulating Toxic Substances

But, we see now through a glass darkly, and the truth, before it is revealed to all, face to face, we see in fragments (alas how illegible) in the error of the world, so we must spell out its faithful signals even when they seem obscure to us. . . .

UMBERTO ECO, *The Name of the Rose*

INTRODUCTION

Assessing Toxic Substances
Through a Glass Darkly

Toxic substances, such as carcinogens,[1] pose threats to human health and well-being. Whether such threats are substantial and our response to them if they are present major scientific and philosophical issues. Yet such substances pose special problems. They are invisible, typically undetectable intruders that can harm us in ways we might not discover for years.[2] We are ignorant for the most part of the mechanisms by which they are transmitted and by which they harm us, which makes their causal path difficult to trace.[3] Many have the potential for catastrophic consequences at least to the affected individuals, yet typically they are associated with modest benefits.[4] The catastrophic injury, though, frequently has a low probability of occurring.[5]

Because of their invisible presence, long latency periods, obscure mechanisms, and untraceability to responsible agents, in assessing the risks from carcinogens we are forced to spell out the "faithful signals" of the fragmentary truth about carcinogens even when "they seem obscure to us." And we must try to detect the signals in a timely fashion before the threats materialize into harm. However, several of the scientific fields that might enable us to discover and accurately assess such risks are in their infancy. Furthermore, the demanding standards of evidence and requirements for certainty commonly considered part of research science frequently frustrate our efforts to identify, regulate, and control toxic substances. A more serious problem is that such procedures are much too slow to evaluate adequately all the substances in commerce or that enter commerce daily.

The need to identify harmful agents in these circumstances may result in mistakes. Mistakes will impose costs on some group or other in the community. However, we must not shrink from this task. Indeed we should expedite our efforts in order to discover the risks and to prevent harm to the people whose lives may be shortened or even ruined because of diseases such as cancer. To achieve this aim, however, the risk assessment and scientific procedures in our legal institutions must be reoriented—we face "paradigm" choices in the use of science for legal purposes. By this I mean that we need to face explicitly how we think about and how we conceive of the use of science in the law. In particular we should avoid the temptation to adopt wholesale the practices of research science into our legal institutions as a number of commentators are recommending, for to do so would frustrate the aims of these institutions. An alternative philosophical approach to science in those areas of the tort and administrative law concerned with environmental health protections would serve the aims of those institutions and the public much better. This book addresses some of these concerns.

The United States in particular and the industrialized world more generally face a potentially serious problem of unknown magnitude: a threat to human health from cancer. In the United States more than 400,000 deaths per year are attributed to cancer.[6] Major sources of these deaths are tobacco (which causes about 25–40% of the deaths due to cancer, with a best estimate in the neighborhood of 30%) and alcohol (which causes about 3–5% of all cancers).[7] Cancer caused by both substances is for the most part self-inflicted harm (except for any effects from inhaling secondary smoke, which now appear substantial).[8] The major way of preventing such harms is for the people involved to reform their behavior and for the industries that actively promote such activities to be curtailed.

However, a number of cancers may be not self-inflicted but caused by others. Conservative estimates suggest that at present about 5–15% of all cancers are caused by workplace exposure (with a best estimate of 10%), about 2% are caused by air pollution, and perhaps as many as 35% are caused by diet.[9] Others estimate environmental causes of cancer much higher. The U.S. Public Health Service estimated that "as much as 20% or more" of cancers now and in the future may result from past exposure to six known carcinogens.[10] Even an industry report indicated that "the full range (of total cancer attributable to occupational exposure) using multiple classifications may be 10 to 33% or perhaps higher, if we had better information on some other potentially carcinogenic substances."[11] This report indicates that asbestos exposure alone may account for 29,000–54,000 cancer deaths.[12] Moreover, even cancer deaths attributed to smoking may hide other causes of cancer because this is a multifactorial process.[13] Several authors have disagreed with the emphasis placed on dietary contributions to cancer.[14] Identifying the environmental causes of cancer is a difficult job with many complications because of the numerous factors that may contribute to the disease.[15] How much of the threat from cancer is created by people as opposed to naturally occurring is not clear. There are at best rough estimates, but the estimates indicated here are sufficiently high to be of concern. If as few as 10–15% of cancer deaths (a very low estimate) are caused by the alterable behavior of others, the number is substantial (40,000–60,000 per year).[16] If the number is greater, as some suggest, the problem is much worse.

If we consider only commercial chemicals as a potential source of cancer,[17] we face considerable uncertainty because we do not know much about their potential toxicity. Of the more than 55,000 chemicals now in commerce, only about 6,000–7,000 have been evaluated in animal tests, many inadequately, with about 10–16% of them testing carcinogenic.[18] Other estimates vary widely. General Electric scientists, examining a seven-volume list from the Public Health Service and using a "relaxed criterion" of carcinogenicity, found that about 80% were carcinogenic.[19] Several federal agencies, reviewing shorter lists of substances about which they may have had some antecedent concerns, found 24–52% of the substances to be carcinogenic.[20]

Against this background it is important to identify carcinogens, to estimate the risks they pose, and then to determine the best policy toward regulating the substances that pose risks of concern. At present, however, little is known about the universe of chemical substances.[21] Few carcinogens have been identified. Moreover, when there are clues to their toxicity, for example, from structural similarity to known carcinogens, these are frequently not followed up.[22] And even when they have been identified in animal tests, only 10%–20% have been evaluated quantitatively for their potency in risk assessments.[23] When such assessments have been completed, the substances have not necessarily been regulated, even though that may be warranted.[24]

One reason (but not the only one) for the delay in the identification, assessment, and regulation of risks from carcinogens is the way scientific evidence is currently used in the legal system. The burdens of proof and other procedures used in science to establish a theory or causal relationship tend to be much more demanding than those adopted in legal institutions. This leads to a conflict between scientific and legal institutional evidentiary norms. If we uncritically adopt scientific standards for legal purposes, we risk frustrating or begging the legal issues.

Human exposure to toxic substances could be and to some extent is regulated by use of the criminal, contract, tort, and administrative law.[25] I focus only on the "regulation" or "control" of toxic substances by means of the tort law and administrative law.

The tort law, also known as personal injury law, establishes public standards of conduct that must be privately enforced by the injured parties or surrogates acting in their name. It aims to provide compensation in order to restore the victim of someone's wrongful activity to the status quo ante, to annul wrongful gains and losses, and to deter conduct that harms others. Thus, the threat of an adverse tort decision is a major way in which the tort law seeks to control exposures to toxic substances.[26] In recent years a number of "toxic tort" suits have been brought to seek recovery for damages from exposures to toxic substances such as asbestos, diethylstilbestrol (DES; an anti-miscarriage drug), paraquat (a pesticide), Agent Orange (a herbicide), and Merrill-Dow's Bendectin (an anti–morning sickness drug). These suits have had substantial effects on the legal system and on the marketing and control of such substances.

Some commentators deplore the current tort law approaches to evaluating scientific evidence and recommend that courts demand better scientific evidence before granting tort recovery. A few courts seem to have adopted similar procedures. This is a mistake, for it appears that such advocates may misunderstand many of the consequences of choosing such a paradigm for torts. Adoption of their recommendations will tend to impose threshold burdens on plaintiffs approaching the criminal law's "proof beyond a reasonable doubt." This will distort the present desirable relationship between plaintiffs and defendants and frustrate some of the major aims of the tort law. We should resist such recommendations. The alternative is to design procedures for accepting scientific evidence in the tort law more in accordance with its goals. Legal and moral norms should determine the adequacy and admissibility of scientific evidence in the law, not vice versa.

Federal and state administrative laws also regulate carcinogens. In the early 1970s, the U.S. Congress passed a number of environmental health statutes aimed to protect human health and the environment from exposure to toxic substances and other pollutants. These laws principally concern the regulation of toxic substances in our food, in our air, in our streams and inland waterways, in our workplaces, and in our drinking water.[27] Two pieces of legislation aim to regulate toxic wastes from "cradle to grave" and to screen the entry of toxic substances into commerce.[28] And one was written to prevent health and environmental problems from arising from abandoned toxic waste sites by requiring their cleanup.[29]

The administrative institutions authorized by these statutes pervade our lives. Yet these agencies are generally less familiar to the public than other areas of the law. They have also received little attention from philosophers, although they have been the objects of study by economists, political scientists, historians, and lawyers. Since these little known institutions are important, consider some of their main features.

First, the administrative or regulatory law[30] is essentially forward looking. Other areas of the law typically provide retrospective remedies—for wrongs done. This is true of tort law. The criminal law typically inflicts loss of liberty, harsh treatment, and condemnation as punishment for wrongdoing,[31] and contract law aims to provide retrospective remedies for breach of contract. Of course, in criminal, tort, and contract law there are prospective deterrent elements, but this seems somewhat less central to these areas of law than the specific future-oriented guidance provided in administrative law.[32]

Second, legislatures (Congress or state legislatures) typically create an administrative agency whose purpose is to protect the community by prescribing specifically the behavior of individuals or firms whose actions may be harmful. The legislation typically authorizes experts to evaluate the appropriate scientific, economic, or other technical problems and to recommend solutions to achieve the goals of the statute.

Third, the agencies are authorized to issue specific rules (regulations) in light of expert advice in order to channel community behavior as part of a comprehensive plan to prevent the harms. Regulations issued under the Resource Conservation and Recovery Act require firms to handle and dispose of their toxic wastes in certain specified ways. The Food, Drug and Cosmetic Act requires firms elaborately to test direct food additives to ensure that they are safe before they can enter commerce. The Clean Air Act authorizes the regulation of air contaminants, if they present threats to our health. Such regulations must measure up to standards specified in the legislation.[33]

Finally, should violations occur despite the agencies' attempts to prevent them, civil and criminal penalties provide retrospective sanctions. These sanctions are designed to ensure compliance with the regulations.

To illustrate features of administrative health law, I frequently use the Occupational Safety and Health Act (the OSH Act) and regulations issued under it. The workplace, consisting of a confined, normally enclosed environment, offers a particularly tractable model for studying environmental health harms, the legislation aimed at preventing them, and the principles that guide the allocation of the costs and benefits of the legislation. Measurements of the toxins are comparatively more accurate in the workplace than in the general environment. And the concentrations of toxic substances are typically much higher in the workplace. This makes the workplace more dangerous to those exposed, but it aids in the control of toxins because exposure-related diseases are easier to detect.[34]

The moral and legal issues may also be somewhat easier, since in workplace environments there is typically a single employer who controls the concentrations of the substances, that is, an identifiable firm or individual to hold responsible. Environmental harms in the wider environment that arise because of the collective actions of hundreds, thousands, or even millions of people pose much more difficult responsibility issues.[35]

Moreover, there has been particularly interesting and influential litigation concerning the Occupational Safety and Health Act and regulations issued under it. One of the most important cases of environmental litigation was the Occupational Safety and Health Administration's (OSHA's) *Benzene* Case, decided by the U.S. Supreme Court in 1980.[36] This case has had considerable influence on the development of environmental law and the scientific practices used in the agencies.[37]

A review of evidentiary procedures in the agencies suggests that we face a philosophical paradigm choice concerning scientific practices here as well. At present regulatory science is fraught with considerable uncertainty and poor evidence for the tasks at hand. Combined with the traditional paradigm of scientific caution and with the standards of evidence scientists typically demand for research purposes, this will frustrate the goals of the environmental health laws. The identification, assessment, and regulation of carcinogenic risks are all slow-moving processes at present. This cannot be attributed solely to the inappropriate use of research paradigms in the agencies—there is plenty of political, economic, and other blame to go around. Nonetheless, a different philosophical approach to regulatory science consistent with the normative aims of the institution would serve those agencies and the public much better.[38]

For both the tort and administrative law to control toxic substances successfully, someone (the plaintiff in a tort suit and typically the government under most regulatory laws) must show that exposure to the substance in question causes disease, or at least increases the risk of disease. However, legal attempts to control toxic substances "stretch" and "stress" the institutions because of inadequate scientific information and the nature of the harm-causing substances. The scientific task is to establish which intruding, invisible molecules are traceable to a source that caused the injuries (or risks) of concern.

To establish the requisite injuries (or risks), scientists may try to estimate the risks indirectly, relying on animal or *in vitro* studies, or to infer the toxic effects on human beings from epidemiological studies. Risk estimations based upon animal studies are plagued by substantial uncertainties, poorly understood biological processes, and few as well as inconclusive data. At the present time risk assessment using such information is not a mature science such as physics, chemistry, or core areas of biology. Moreover, present risk assessment methods are time-consuming, labor intensive, and costly, thus preventing more rapid identification, assessment, and ultimately regulation of carcinogenic risks.

Epidemiological studies are both frequently insensitive and plagued by numerous practical problems; either may prevent the detection of risks of concern. And in many cases, but not necessarily all, such studies must sacrifice either scientific accuracy or evidentiary sensitivity—one cannot have both.

Because of these problems, then, we can glimpse the truth about harms posed by carcinogenic substances "in fragments (alas how illegible) in the error of the world," but we are forced by the necessities of regulation and tort law suit settlement "to spell out their faithful signals even when they seem obscure to us." Scientists, facing considerable uncertainty and time constraints, cannot wait until all the data are in to draw their conclusions. If they do, people may suffer disease and death as a consequence. There may be mistakes from decisions to take precautionary regulatory action or a favorable tort law decision for a plaintiff when the best data are not available, as well as mistakes from a decision not to regulate or not to permit compensation.

One kind of mistake (a false positive) occurs when a substance is wrongly thought to cause harm or risks of harm but in fact it does not. A similar but less serious mistake is that the assessment procedures can lead to "overregulation"—greater regulation of the substance under the applicable law than is warranted by the harm it does. In the tort law such mistakes can lead to overcompensation of plaintiffs.

A mistake in the other direction (a false negative) results when a substance is wrongly believed to be "safe" when it is not. A lesser error is "underregulation"—lesser regulation of the substance under the appropriate law than is justified by the harm it causes. The analogous mistake in torts results in undercompensation of plaintiffs.

Mistakes impose costs on someone: false positives, overregulation, and overcompensation of plaintiffs impose costs on manufacturers of the substance, on their shareholders, and on the consumers of their products. We may be deprived of useful products or pay higher prices for them. False negatives, underregulation, and undercompensation of plaintiffs impose costs on the victims or on those put at risk from the toxicity of the substance. On whom the costs of such mistakes should fall is a normative question of substantial moment. The very existence of firms or their product lines or the welfare of the public may be threatened on the one hand.[39] Human death, disease, and compromised quality of life and their associated economic costs may result on the other.

The number and kinds of mistakes that are made will depend upon how the risk estimation tools of animal bioassays and epidemiological studies are used for legal purposes. Traditional scientific practices in using these tools, practices typical of research science where the discovery of truth for its own sake is the primary aim, in many cases will beg or frustrate some of the normative goals of the law. Risk assessments relying upon animal studies and aspiring to the goals typical of research practices will paralyze regulation and thwart the health-protective aims of the environmental health law. Use of traditional demanding standards of scientific evidence to protect against false positives may well beg the normative issues in question before a regulatory agency or a tort court. The result might be studies that are too insensitive to detect risks of concern or too expensive to conduct (when they are appropriately sensitive); both can easily leave people at risk.

Many of the shortcomings in the use of risk estimation procedures can be addressed. However, remedies for the problems will require a different approach—a change in philosophy, in our paradigms—in the use of science for legal purposes.

Failure to modify our approaches in these institutions will likely leave us at greater risk from toxic substances than we otherwise would be. Such choices can occur and be justified, legally and morally, if we appreciate more of the social consequences of using scientific practices in their legal settings. Thus, the approach to assessing evidence about toxic substances must be guided not so much by the goals of research science, where the pursuit of truth is the main or sole aim, but by the appropriate legal norms of the institutions in which the information will be used. A number of commentators, some aiming to protect public health (and some less motivated by this concern), recommend scientific approaches which in fact will exacerbate slow regulation and false negatives, not improve the situation.

To address such pressing issues we must understand the legal institutions and their aims and limits, as well as the scientific tools available for identifying and assessing the risks from carcinogens and their role in the institutions. However, we must face the use and interpretation of scientific evidence in these contexts as normative matters, much as we have in designing legal procedures, so that our evidentiary procedures promote and do not frustrate our legal goals. We face philosophical choices in how we think about scientific evidence in such forums.

These issues raise important philosophical questions about the institutions involved and about the evidentiary procedures used to establish the causal claims needed in torts or in administrative agencies. Although philosophers have long discussed issues in philosophy of law, traditionally they have tended to focus on the criminal, tort (to some extent), and property law.[40] More recently they have considered topics in torts and in contract law.[41] Administrative law, however, has received little philosophical discussion, even though in today's industrial and technological society it has been developed to cope in part with environmental health problems and it pervades our lives.

This book is an attempt partially to rectify this shortcoming by focusing on an issue that is important to the public health and national well-being. It is an essay in risk assessment and the philosophy of law, an essay about the use of science in the law concerning toxic substances. It focuses on aspects of the scientific and legal problems posed by the attempt to control human exposure to toxic substances and on some philosophical approaches to addressing these problems within tort and administrative law. It is not a wholesale evaluation or critique of either institution. Instead it addresses only the use of risk estimation procedures within these legal settings and how our use of such procedures is likely to affect the legal institutions.

Chapter 1 provides some of the scientific background of risk assessment and sets up some of the institutional questions concerning the appropriate evidentiary standards needed within torts and administrative law to establish the requisite harms or risks of harm.

Chapter 2 considers the use of scientific evidence in the tort law. Recent proposals, seeking to require more demanding scientific evidence in toxic tort (and other) suits, are mistaken in their choice of paradigm. The appropriate paradigm is not that for scientific practice but the traditional tort law standard of evidence articulated in *Ferebee v. Chevron Chemical Co.*, which says the appropriate standard is not scientific certainty but "legal sufficiency."[42] The paradigm of choice in torts is in a sense to retain much of the status quo. We should *not adopt* for purposes of tort law compensation the standards of research science, as some commentators have recommended and as some courts have held.

Chapter 3 addresses a problem that infects both regulatory and tort law efforts to control exposure to carcinogens. Diseases can be causally overdetermined; that is, there are two or more possible causes, each of which alone would be sufficient to explain the presence of disease. What should be done when a disease may be the result of such "joint causation"? The tort law has developed a solution to this problem which has recently been adopted by some courts considering toxic torts. However, the problem of joint causation and more general issues raised by causation reveal weaknesses in the tort law for compensating victims exposed to toxic substances. Because of these and other problems with the tort law, administrative institutions at least *in principle* may better control human exposure to toxic substances. However, such institutions face enough practical problems that they will not always reliably protect us from toxic substances.

Chapter 4 discusses some of the risk assessment issues that arise in administrative agencies charged with regulating carcinogens. I argue that present assessment strategies, as well as some recommended by commentators, both of which are temptingly inspired by the paradigm of research science—the use of careful, detailed, science-intensive, substance-by-substance risk assessments—paralyze regulation. Alternative approaches to regulatory science which acknowledge and use normative considerations to guide assessment procedures and which recognize the importance of the rate of evaluation will expedite risk assessments and reduce regulatory false negatives and underregulation. In this case I suggest a paradigm shift away from the norms of research science to presently available quicker but reliable approximation procedures for achieving much the same results. This is a recommendation away from the status quo in the agencies and away from current trends in risk assessment, which emphasize more science-intensive approaches. The use of approximation procedures has a long and honorable tradition in science to simplify calculations and to expedite problem solving; they are surely appropriate in regulatory institutions where the rate of assessment is important. The regulatory challenge given current problems is to utilize faster assessments that are presently available. The scientific challenge is to refine existing procedures and to develop others to expedite the identification and assessment of carcinogens.

Moreover, although some of the suggestions about regulatory science address current imperfections in scientific knowledge and understanding, not all do. That is, even if we had perfect scientific understanding of the biological mechanisms of toxic substances, using it could still be time-consuming and labor intensive when applied in the regulatory setting. If it is, the public will continue to bear the costs of inaction and sluggish identification, assessment, and regulatory processes through their morbidity and morality. Thus, even with perfect scientific knowledge, the scientific and regulatory challenge will be to refine expedited approximation procedures to identify and assess toxins, so the costs of scientific investigations are not inadvertently imposed on the public.

The choice of evidentiary standards in tort and administrative law is in effect a choice between imposing the costs of overregulation and overcompensation on individuals and the public and imposing the costs of underregulation and undercompensation—mainly adverse health effects on individuals. Chapter 5 provides some justification for both the epistemic and moral presuppositions for much of the argument of the book. Evidentiary standards should be chosen relative to the institution in which they are used. This is because institutional norms indicate the importance or urgency of avoiding different kinds of epistemic mistakes. In addition, most of my criticism of scientific practices recommended for the tort or administrative law is that they incur or risk too many false negatives. Thus, I consider several moral theories in order to evaluate their strengths and weaknesses for assigning importance or urgency to both false negatives and false positives. Although I do not defend with great specificity one moral view to address such issues, I argue that justice and other principles sensitive to the effects of policies on individual persons (so-called distributively sensitive principles) require that priority be given to avoiding false negatives and underregulation.

Because our present lack of scientific knowledge will probably extend for some time into the future, we are condemned to assessing and regulating toxic substances "through a glass darkly." We must recognize the normative implications and components of our risk assessment procedures. We must also design those procedures to be more expeditious, especially in the agencies, or we will continue to be frustrated in our efforts to identify, evaluate, and control the large number of potentially toxic substances that are already in commerce (with three to five added every day). A clear moral view will not clarify the glass through which we evaluate potentially toxic substances, but it may remind us of the fundamental aims of the tort and regulatory law, and it may give us greater appreciation of some of the costs of our scientific procedures. It can guide us when the facts are so uncertain, because, when we have to make decisions in the absence of good information, we must rely more heavily on the values we seek to foster. Finally, in reminding us of what is at stake, it might even motivate us to adopt evidentiary procedures that better serve our goals than at present.

1

The Scientific Background

Federal and state administrative agencies use scientific or quasi-scientific "risk assessment" procedures to provide evidence about risks from toxic substances. The aim of agency actions is to estimate the risks to human beings from exposure to toxins in order to prevent or reduce those risks. In the tort law a plaintiff has to establish a causal connection between a defendant's alleged faulty action and the plaintiff's harm.

This chapter considers animal bioassays and human epidemiological studies, two aspects of carcinogen risk assessment relied upon in these institutions to ascertain risks to human beings.[1] The best evidence that a substance causes cancer to human beings is provided by *well-done* epidemiological studies with large samples and sufficient follow-up. However, I begin by considering animal studies. This is the evidence much more frequently relied upon by regulatory agencies, although it is less typically used in the tort law and some jurisdictions give it little credence.[2]

The uncertainties inherent in inferences from animal bioassays present one class of problems in their use for the predictions of risks to human beings. Both actual and possible scientific uncertainties in projecting risks to human beings are large enough that two different researchers using exactly the same data points from an animal study can come to much different conclusions. The uncertainties and the policies used to overcome them permeate regulatory science with public policy or moral considerations. As a result, regulatory scientific decisions are mixed judgments of science and value, and quite properly so. This makes regulatory science much more normative and much less like ordinary, core areas of science than we might suppose. Consequently, although many assume that risk assessment is independent of risk management decisions, in the present state of knowledge this seems mistaken.

Moreover, if the procedures of research science are transposed to regulatory contexts, they may determine and even beg the public policy outcomes in ways often unbeknownst to practitioners. Scientists unwisely demanding more and better data, withholding scientific judgment until there is sufficient research, and using too demanding standards of evidence can frustrate legal aims.[3] Taken individually each of the foregoing may frustrate the discovery of risks of concern for the substance under consideration. Perhaps of greater importance, the combined effect of these tendencies is to slow the scientific evaluation of carcinogens, and thus regulatory efforts, to a snail's pace. This prevents agencies from assessing identified but unevaluated carcinogens or from devoting resources to identifying other toxins. The remedy considered in Chapter 4 is to develop scientific and legal approaches which minimize these problems and better serve institutional goals.

Epidemiological studies pose a different set of problems. In many circumstances unwitting commitment to traditional scientific procedures in the design and interpretation of epidemiological studies (and scientific data more generally) can beg the normative issues at stake. Even when there are no practical evidence-gathering problems or none of the uncertainties noted previously, either scientists or the risk managers who use their data may be forced into a dilemma: they may have to choose between adhering to the evidentiary standards typical of research science and interpreting them in "regulatorily sensitive" ways— providing results that are sensitive enough to detect the risks of concern and avoiding false negatives.

Risk assessments based upon animal and human studies suffer from shortcomings that can easily frustrate the aims of the legal institutions in which they are used and thus affect the regulation of toxic substances. Whether these problems arise and whether they are exacerbated or ameliorated depends upon the scientific and legal responses to them. A tempting response is to make risk assessments more nearly like normal science. In many circumstances this is very likely to increase the risks to human beings. Consequently, as I argue throughout, decision makers should balance the pursuit of the goals of research science with those of the legal institutions in which they are used.

PREDICTING RISKS FROM ANIMAL BIOASSAYS

Background

Risk assessment aims at providing accurate information about risks to human beings so that agencies in fulfillment of their statutory mandates can regulate exposure to potentially carcinogenic substances. After scientists in the technical, scientific part of the federal agencies have provided an estimate of risks to human beings from exposure to toxic substances, they then give this information to the risk managers. *Risk management*[4] is concerned with managing the risks in accordance with statutory requirements and other economic, political, and normative considerations.

Risk assessment can be divided into hazard identification, dose–response assessment, and environmental risk assessment. Hazard identification is "the process of determining whether exposure to an agent can cause an increase in the incidence of a health condition (cancer, birth defect, etc.)."[5] Dose–response assessment seeks to characterize quantitatively "the relation between the dose of an agent administered or received and the incidence of an adverse health effect in exposed populations and estimating the incidence of the effect as a function of human exposure to the agent."[6] Both hazard identification and dose–response assessment rely on inferences from animal bioassays and human epidemiological studies. Exposure assessment estimates risks to human beings from exposure to carcinogens when they are released into the environment, soil, groundwater, and air. I focus mainly on hazard identification and dose–response assessment.

We should recognize at the outset, however, that risk assessment in the present state of knowledge is a third-best solution to the problem of estimating harms to human beings from exposure to toxic substances. The ideal is "harm assessment"; if we had perfect information, we would accurately assess the harmful effects to people and the environment from exposure to toxic substances. This would provide us with exact numbers of deaths, diseases, and environmental harms, and we would not overestimate or underestimate the effects of toxic exposures.[7]

If we distinguish between risk and uncertainties, a risk is the *probability* of an unfortunate or undesirable outcome,[8] when such probabilities can be assigned to outcomes. Thus, a "risk assessment" properly speaking aims to estimate the probabilities of harms from toxic exposures and is a second-best solution to a harm assessment. For a whole population and for accurate probabilities of harm, this alternative would very closely approximate the morbidity and mortality rates of a harm assessment.

At present the task of regulators is more complicated than this, for great uncertainties can obtain in trying to predict such harms. For example, one can show mathematically that the projection of risks to human beings can vary by several orders of magnitude depending upon the choice of high-dose to low-dose extrapolation models.[9] Some believe there is little biological basis for choosing between these models,[10] although even on this point there is disagreement.[11] Thus, we should think of risk assessments not as risk assessments properly speaking, but as "risk and uncertainty assessments." This is the third-best solution to a harm assessment. For my purposes "risk assessment" will refer to this third possibility—the present state of the art.

Matters of considerable moment depend upon the products of risk assessment, for in many cases one answer (a projection of high enough risks to require regulation) may impose substantial costs on the affected industry and perhaps the larger public. On the other hand, another answer (a projection of a risk low enough so that regulation is not required) may leave innocent people at risk from exposure to dangerous substances.

Because of the uncertainties in risk assessments, regulatory agencies will make mistakes. Two kinds of mistakes can be made at the level of hazard identification: false positives and false negatives. A false positive occurs when one mistakenly identifies a substance as a carcinogen. A false negative is a failure to identify a substance as a carcinogen when it is one.

A second class of mistakes might occur at the level of dose–response assessment and regulation: overregulation or underregulation of the substances in question. Overregulation occurs when a substance is regulated in accordance with a particular statute too stringently for the kind and degree of harm that it causes. The substance might cause no harm of regulatory concern, or much less harm than an agency believed. By contrast, underregulation occurs when a substance is regulated under a particular statute to a much lesser degree than it should be. In what follows often I use "false positives" to refer generically to the mistaken identification of a substance as harmful and to overregulation. Similarly, "false negatives" often refers generically to a failure to identify toxins and underregulation. (When it is necessary to distinquish between mistakes of identification and regulation, I do so.) Both kinds of mistakes are illustrated in Table 1–1.

The terms "false positive" and "false negative" have been borrowed from science (statistics in particular) and adapted to make this more general point. Because probabilities are involved in statistical studies, it is almost certain that if one

Table 1–1 False Positives and False Negatives

Possible Test Results	Possibilities in the Real World of Causal Relationships	
	Null hypothesis is true: Benzene *is not* positively associated with leukemia	Null hypothesis is false: Benzene exposure *is* positively associated with leukemia
Test does not show that benzene exposure is associated with leukemia	No error	False negative
Test shows that benzene exposure *is* associated with leukemia	False positive	No error

were regulating large numbers of substances, there would be both false positives and false negatives, for by chance alone mistakes would be made. By analogy it is likely that agencies regulating large numbers of substances will make mistakes, either as a consequence of the underlying statistical studies or through other errors. The possibility of regulatory mistakes raises the normative question of how to cope with the uncertainties in risk assessments: On whom should the costs of regulation or its absence fall? We return to these points throughout the book and focus on their distribution specifically in Chapter 5.

Regulatory Science and Policy Choices

The primary method for estimating cancer risks to human beings in the regulatory setting is to study the carcinogenic effects of substances on animals and then to project risks to human beings based upon this information. In animal studies, three or four small experimental groups of rodents are fed high doses of a substance to see whether the tumor rate in the experimental groups is significantly greater than the cancer rate in a control group. If it is, then scientists extrapolate from response rates to high-dose exposures in rodents to project response rates at low-dose exposures in rodents (an exposure rate much closer to the typical human exposure dose). Risk assessors use this low-dose response rate along with principles of biology, toxicology, and pharmacology (if this information is used) to estimate, on the basis of rodent to human models, a dose–response function, the likely risks that human beings would face at hypothetical levels of exposure. This risk information is then combined with actual exposure information at all doses on the appropriate population at risk in order to estimate the magnitude and extent of the risk to human beings. Finally, the risk information is combined with economic, policy, statutory, and technological feasibility information so that regulatory agencies can then decide how best to manage properly the risks in question. Figure 1–1 is a schematic of the risk assessment–risk management relationship as it is presently conceived.

There are a number of advantages to using animal studies as evidence that a

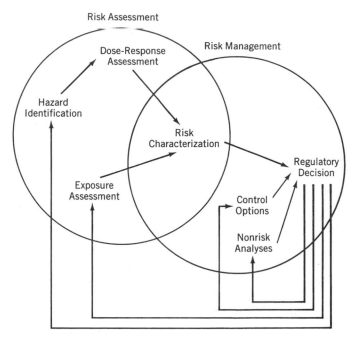

Figure 1–1   Relation between risk assessment and risk management. (From the U.S. Environmental Protection Agency.) Risk Communication Training Documents

substance causes cancer in human beings. Many experts believe that "animal evidence alone should serve as the basis" for regulating carcinogens.[12] Most substances that induce cancer in one mammalian species also induce cancer in others. A finding of "carcinogenicity in rodents is proof that the chemical is carcinogenic in a mammalian species."[13] The pathological development of tumors in various species of animals in most cases is believed to resemble that in humans. Human and animal molecular, cellular, tissue, and organ functions are thought to be similar.[14] "In the absence of adequate data on humans, it is reasonable, for practical purposes, to regard chemicals for which there is sufficient evidence of carcinogenicity in animals *as if they presented* a carcinogenic risk to humans."[15]

Animal studies also have several advantages over human epidemiological studies. For one thing, few industrial chemicals have been adequately tested by epidemiological studies to discover whether they cause cancer in humans. (As of 1988 researchers seeking to compare epidemiological with animal bioassay estimates of cancer potency considered epidemiological results for only 23 chemicals.[16]) For another, epidemiological studies are frequently too insensitive to detect relative risks of concern. Further, an epidemiological survey that is not positive is of questionable merit for showing a substance is "safe" because of sensitivity problems. (Both of these points are considered later.) Moreover, it may still be too soon to observe the carcinogenic effects of many substances, since cancer typically has latency periods of up to 40 years.[17] Even after a sufficient latency period has elapsed, it may be difficult to trace diseases to particular substances because of the insensitivity of epidemiological studies and because almost no toxic substances leave a unique "fingerprint" of their presence.[18] Animal studies are faster and cheaper than human studies. Moral considerations also provide reasons for using animal studies; there is no justification to wait for "evidence of harm in exposed people when risks can be established relatively quickly by animal experimentation."[19]

Animal studies have some shortcomings, however, which we should understand in order to appreciate the foundation of regulatory standard setting and some of their evidentiary limitations. Estimating risks from animal studies requires a number of inferences from the established experimental data from laboratory animals to the projection of end point risks to human beings. These inferences have a number of uncertainties and inference gaps that must be bridged in order to produce the risk numbers. Inference gaps arise because there is *insufficient information* (in both theories and data) available to settle the scientific questions at issue. These gaps are distinguished from *measurement uncertainties*, which are features of scientific information that cannot be *measured* precisely. Rather, in many cases the inference gaps result from insufficient understanding of the biological mechanisms involved or the relationships between biological effects on one species compared to another. Scientists typically use mathematical models or other generalizations to fill the gaps, and these might produce scientific predictions that differ substantially from one another depending upon the model chosen. Thus the range of possible answers will produce a range of uncertainty. Such models introduce uncertainties because surrogates for the proper quantities (if they have even been identified) are used, because some possibly appropriate variables may be excluded, or because the proper model for representing and quantifying the data is not known.[20] Moreover, uncertainties can be introduced because of the inherent variability in biological response between individuals.[21] Concerns about the uncertainties in arriving at dose–response (potency) estimates for substances lead to considerable controversy among critics about their use in predicting risks to human beings.[22] Nonetheless, there is a substantial consensus for continuing their use.[23] To see these points consider a few of the detailed steps in risk assessment procedures.

In an ideal animal study, there are usually three or sometimes four groups of about 50 animals each that are studied: (1) a control group; (2) an experimental group fed the maximum tolerated dose (MTD) of a toxic substance, "a dose as high as possible without shortening the animals' lives from noncarcinogenic toxic effects;"[24] (3) an experimental group fed one-half the MTD; and (4) sometimes an experimental group fed one-fourth the MTD. Tumor data from each of the experimental groups then become fixed data points on a graph, if the tumor results differ statistically from the control group. (This is schematically indicated in Figure 1–2.) To estimate low-dose responses in animals, researchers then use a computer model to fit a mathematical curve through experimental data points and through those seen in control groups. (Figures 1–3 through 1–5 indicate the general problem.) Figure 1–5 is misleading in some respects. This is an extrapolation from the same animal data using different mathematical models, and they can extrapolate to quite different values. However, there are biological constraints such that some of these extrapolations do not make sense. (These will be discussed later.) Moreover, for all the

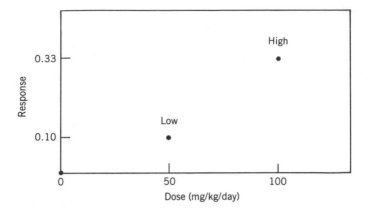

Figure 1–2 Dose–response evaluation performed to estimate the incidence of the adverse effect as a function of the magnitude of human exposure to a substance.

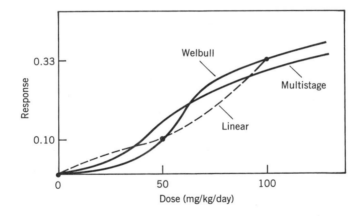

Figure 1–3 Dose–response curve. (From the Environmental Protection Agency.) Risk Communication Training Documents.

models, if the background cancer in control animals (or the general population) works by the same mechanism as cancer induced by the administered carcinogen, resulting in an additive carcinogenic effect, all the models become linear at low doses, rather than curvilinear as they appear here. As long as the background cancer rate is 1% or higher, this linear effect results. In addition, all the curves are shifted to the left because of the additivity effect.[25]

One kind of uncertainty is introduced because the etiology of carcinogenesis is insufficiently understood to enable scientists to know which mathematical model is the correct one for representing the mechanism of carcinogenesis and for making the extrapolation from high-dose effects in animals (i.e., from the experimentally established data points) to low-dose effects in animals (i.e., to effects typically beyond the experimentally established data points).[26] In addition, there are insufficient data

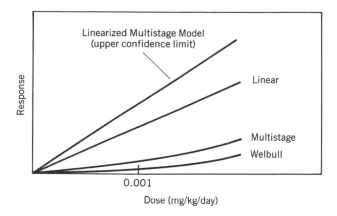

Figure 1–4 High-dose to low-dose problem.

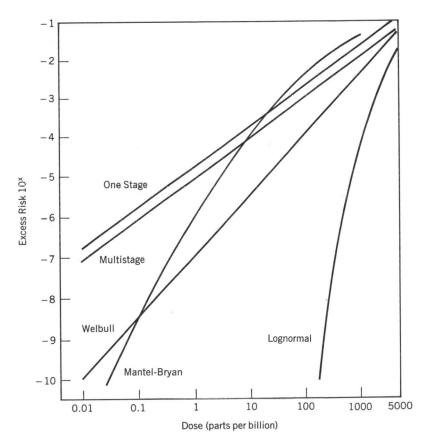

Figure 1–5 Extrapolated risks due to benzene exposure.

points based upon experimental evidence to enable researchers to find a unique curve to fit the data produced by controlled experiments.[27] Thus, both insufficient theoreti cal understanding and too few data contribute to the uncertainties that plague dose– response risk assessment procedures.

After extrapolating from high-dose effects to low-dose effects (more typical of human exposure) in animals, risk assessors must estimate the low-dose effects on human beings. The use of different mouse (or rodent) to human extrapolation models can produce differences in risk estimates—estimates that range from a factor of 6 to 13 to 35, depending upon which of several methods is used.[28] In this case, the difference is due partly to differences in theoretical understanding and partly to lack of a sufficient experimental data base for toxic substances.

Other uncertainties that plague toxicological risk assessment concern the relative weighting of positive and negative animal studies for the same substance, whether benign and malignant tumors should be weighted equally as evidence for carcino- genicity in animals (and thus in humans), whether and how data from two or more nonhuman species should be combined, and so on, but I do not pursue those here.[29] Some of these sources of uncertainty are summarized in Appendix A.

The foregoing steps are some of the main ones leading to an estimate of the carcinogenic potency of a substance. The next step is to indicate how much of a substance will reach human beings through the environment—via soil, air, water, or food. Environmental fate models provide a way of estimating this. There are also uncertainties in these procedures. For example, models for predicting the transport of ethylbenzene from leaking gasoline tanks differ by as much as a factor of 1500, and there appears no easy way to validate the correct model for making such assessments.[30] Lee and Chang point out that although theoretical models for predict- ing the environmental fate of substances such as benzene, xylene, toluene, and ethylbenzene are reasonably well understood at least in laboratory settings, applying the models under actual field conditions could produce substantial differences in the estimates of the risks.[31] The differences in this case are for the most part due not to a failure of theoretical understanding of the environmental fate of substances under controlled conditions, for this seems to be reasonably well understood. Instead there is typically insufficiently detailed site-specific information about the behavior of substances in the soil, groundwater, or air.[32] Additional on-site field research could remove some of these uncertainties (this is not true for many of the uncertainties connected with toxicological risk assessment), but it is time-consuming, probably expensive, and of limited value since the results could vary by individual substance, type of soil, underlying impermeable stratum, groundwater properties, and other factors specific to the location.

The cumulative theoretical uncertainties that can be introduced by uncertainties at each step of the risk assessment process could be substantial. For example, if, at each step of a chain of inferences, alternative inference guidelines or choice of models would introduce differences in a risk assessment of only magnitude 2, and if there were 10 such inference gaps, then the cumulative theoretical differences math- ematically could be as large as 1024 ($2^{10} = 1024$). If there were 20 such gaps where each choice of a different model would only make a difference of 2, it is math- ematically possible for the cumulative difference to be as large as 1,048,576

$(2^{20} = 1,048,576)$.[33] [Even though these are mathematical possibilities, there are not typically so many gaps and such large differences do not materialize (discussed later).]

The potential quantitative differences between models to fill an inference gap in a carcinogen risk assessment, however, could easily be much greater than a factor of 2. In toxicological risk assessment, high-dose to low-dose extrapolation models can vary by several orders of magnitude.[34] The use of upper confidence limits versus maximum likelihood estimates (a very unstable point)[35] in estimating high-dose to low-dose extrapolations can vary from a factor of 2–5 where there are good dose–response data, up to several orders of magnitude at the lowest doses, where there are not. Interspecies scaling factors, used to account for the different toxicological effect in different mammalian species, can vary up to a factor of 35. Use of pharmacokinetic information, which enables a scientist to estimate the dose of a substance reaching an internal target organ, may change a risk assessment by a factor of 5 to 10 or even more compared with the dose of a substance reaching an external exchange boundary in an animal or human, such as the nose, skin, or mouth.[36] An agency's choice of *de minimis* risk thresholds, which trigger regulatory action, might differ by one or two orders of magnitude from that of another agency.[37]

However, the logically possible uncertainties that might exist in risk assessment have not tended to materialize in actual risk assessments between agencies or other groups performing them. For one thing, for many risk assessments, there may not be as many inference gaps as the possibilities sometimes suggest. For another, empirical data constrain some of the choices. For example, some high-dose to low-dose extrapolation models "seem impossible to interpret in terms of any biological description."[38] Moreover, where there are good data on both animal and human response rates many of the typical assumptions used to infer human risks from animal studies tend to predict risks to humans that agree fairly closely with epidemiological studies.[39] Furthermore, different regulatory agencies frequently agree on the same models, even though scientific data and theories are not fully adequate to support such choices. And there has been some attempt at coordination between federal and state agencies.

While the differences between actual risk assessments have tended not to be as great as the mathematical possibilities, there still can be substantial discrepancies in risk assessments for particular substances. Some of these results for two agencies are summarized in Figure 1–6 and Appendix B. A comparison of potency estimates done by the California Department of Health Services and by the U.S. Environmental Protection Agency's Carcinogen Assessment Group shows that the California Department of Health Services projected a higher potency and thus a higher risk for 17 of 27 carcinogens examined, while EPA projected a lower potency for 10 of the 27. Where the California Department of Health Services projected a higher number it tended to be from 1.1 to 20 times higher than the Environmental Protection Agency. When the California Department of Health Services projected a lower number than the EPA, the range was from 0.7 to 0.006 (or three orders of magnitude).[40] Although the extreme discrepancies are the exception to fairly close agreement between the agencies, they do show that there can be major disagreements between them nonetheless.

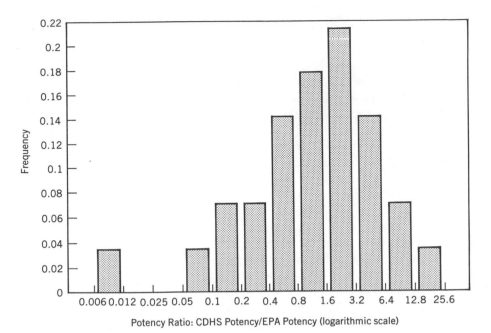

Figure 1–6   Frequency diagram of differences between EPA and CDHS potency values. [From Lauren Zeise, "Issues in State Risk Assessment: California Department of Health Services," in *Proceedings: Pesticides and Other Toxics: Assessing Their Risks,* ed. Janet L. White (University of California, Riverside, College of Natural and Agricultural Sciences, 1990), p. 138.]

The point of the preceding remarks is that carcinogen risk assessments

1. can suffer from substantial uncertainties and
2. in fact sometimes can differ sufficiently to make a substantial difference in the actual risks that are predicted, and
3. may well in some cases make substantial differences in regulations that are issued.

In order to provide guidance for the gaps created by the absence of data and theoretical understanding, a National Academy of Sciences report suggested that for the approximately 50 inference gaps that exist in the procedures for estimating risks to human beings from basic toxicological information regulatory agencies adopt policy, or inference guidelines.[41] These inference guidelines constitute assumptions to bridge the gaps in question.[42] Agencies have adopted four kinds of assumptions:

1. assumptions used when data are not available in a particular case;
2. assumptions potentially testable, but not yet tested;
3. assumptions that probably cannot be tested because of experimental limitations; and
4. assumptions that cannot be tested because of ethical considerations.[43]

Such assumptions guide many of the choices of model mentioned earlier.

The policies adopted to bridge such gaps can lead to considerable controversy, for they can make a difference in the estimation of risks to human beings. Frequently, little in the way of biological evidence in the present state of knowledge can determine the choice between the models.

I believe, however, that we cannot avoid the use of inference guidelines, since we need to make decisions concerning the effects of toxic substances on human beings. Part of the reason is that the scientific basis of risk assessment, because of the uncertainties involved, exhibits *in a much more radical form* a feature of all scientific inferences (all empirical inferences for that matter): the evidence available for the inference *underdetermines* the inference or the theory that aims to explain the evidence. This general point about our empirical theories and beliefs was well articulated by Willard Van Orman Quine in *Word and Object*:

> To the same degree that the radical translation of sentences is under-determined by the totality of dispositions to verbal behavior, our own theories and beliefs in general are under-determined by the totality of possible sensory evidence . . .[44]

Although there has been considerable discussion pro and con of Quine's thesis, especially that concerning the problem of radical translation, there is ample support for his general point that evidence underdetermines theories or scientific inferences. First, at least since Descartes philosophers have discussed the possibility of persons having mistaken beliefs based upon the evidence before them. The possibility of mistakes (in the extreme cases imagined by Descartes) indicates that the evidence does not *guarantee* the conclusions one might typically infer from it.[45] Second, a paradigm of scientific inference, so-called inference to the best explanation, rests on the possibility that there may well be several plausible alternative explanations to account for the available evidence to an observer. The resulting problem is to characterize which explanation is the best one. Again, the possibility (and in some cases the likelihood) of alternative explanations in science is evidence for the general thesis of underdetermination.

The point of the preceding remarks is that the evidentiary underdetermination of an inference is both an old problem for all empirical beliefs (Descartes's point) and a problem common to all scientific inferences (Quine and the point about inference to the best explanation). Thus, in carcinogen risk assessment the fact that the available evidence does not uniquely determine a correct model to characterize the mechanisms or many of the other aspects of scientific behavior for each substance is not *new*. However, the underdetermination of models or theories in carcinogen risk assessment is so much more *radical* in carcinogen risk assessments as to make it substantially different from the case of ordinary scientific inferences. The uncertainties resulting from lack of understanding are much greater than in more well-established areas of science.

The last point, then, leads to the following additional points: inference guidelines, or at least some kind of choices between competing models, cannot be avoided, if risk assessors are going to provide needed evidence for the assessment and regulation of toxic substances. The inference guidelines (or choices) must be chosen (made) on the basis of some reason, but since scientific data radically underdetermine the choice and perhaps more radically underdetermine this choice than similar choices in any other areas of science, some *other* consideration(s) must determine it.

These additional considerations are typically of several kinds. Some are *scientific* or other *empirical* generalizations, but ones not necessarily well supported in a particular case (such as the claim adopted by many agencies that carcinogens do not act by means of a threshold mechanism, or the claim that since a surface area extrapolation is used for other purposes of comparing the metabolic activity between species, a similar extrapolation model should be used to predict the toxic effects of a substance from one substance to another). Sometimes decisions are made on the basis of *simplicity* or *ease of calculation*.[46] And agencies sometimes choose a middle-ground theory simply because it is *midway between alternatives* they might adopt.[47]

In addition, however, nearly all agencies acknowledge that some of the decisions of scientific models are chosen on *nonscientific policy or moral grounds*. Typically they reason that risk assessment procedures should not underestimate risks to human beings at the end of the process or that they must be prudent in protecting public health. They choose not to underestimate the risks to human beings by using a health-protective high-dose to low-dose extrapolation model, by using the biological response in the most tumor-sensitive mammalian species as a basis of risk assessment, and by counting benign as well as malignant tumors as evidence of a substance causing a carcinogenic response in experimental animals. Even though agencies acknowledge the policy role in risk assessment, it is not clear that all their choices are health protective, for they ignore other choices that could reduce the underestimation of risks. For example, sensitive subpopulations, such as the elderly or nursing infants, are seldom considered. In some cases humans are much more sensitive to substances than test animals (e.g., benzidine), and the timing of exposure is frequently not taken into account.[48]

The National Academy of Sciences argued that important inference guidelines are based upon policy considerations:

> Guidelines unavoidably embody both scientific knowledge and risk assessment policy. In the past, regulatory agencies typically used a conservative [health-protective] approach in the development of risk assessment policy, as in the choice of the most sensitive species, use of the most conservative dose–response curve, and the lack of acceptance of negative epidemiologic data.[49]

The Academy report notes that risk assessments *could* avoid the use of policy guidelines, but argues that

> guidelines very different from the kinds described could be designed to be devoid of risk assessment policy choices. They would state the scientifically plausible inference options for each risk assessment component without attempting to select or even suggest a preferred inference option. However, a risk assessment based on such guidelines (containing all the plausible options for perhaps 40 components) could result in such a wide range of risk estimates that the analysis would not be useful to a regulator or to the public. Furthermore, regulators could reach conclusions based on the ad hoc exercise of risk assessment policy decisions.[50]

Ad hoc choices between plausible options likely would still involve policy considerations, but they would be hidden from the public, not identified as public choices
on the part of an agency.

These considerations do not deductively warrant the conclusion that risk assessments in the present state of knowledge are policy or value or morally laden, but
given current practice and the radical underdetermination of the theory by the evidence, they strongly support such conclusions. Thus, we might somewhat oversimplify the point by saying that whether a risk assessment indicates a risk to human
beings and the seriousness of that risk are in part, normative policy considerations.
Thus, not just the notion of an *acceptable risk* (an obvious normative concept), but
(in the present state of knowledge of carcinogen risk assessment) also the ideas of *a
risk* and the *severity of the risk* are partly normative notions, for they are the product
of normative judgments.[51]

Given the uncertainties in the quasi-scientific field of risk assessment and given
the typical procedures adopted by agencies for coping with such uncertainties
(and the recommendations by the National Academy of Sciences for coping with
them), the scientific neutrality of factual inquiries that appear to be typical of normal
scientific inquiries (in subatomic physics, molecular biology, geology, etc.) does not
obtain for risk assessment. This point is very clear if one focuses on a particular risk
that is the outcome of the risk assessment process or on agency risk assessment
guidelines.

The upshot of the foregoing observations and arguments is the following: *in fact*
actual risk assessments have relied heavily upon policy or moral considerations, *in
fact* agencies at present rely upon such considerations, and the National Academy of
Sciences *recommends* that agencies continue to rely upon such considerations. Finally, because of these arguments it seems *proper and plausible* that agencies should
continue to do so (I return to this point in Chapter 4). Thus, it appears the quasi-
scientific process of risk assessment in its present state is substantially different from
other areas of science; it is a process substantially permeated by policy considerations and seems destined to continue to be so permeated into the indefinite future.
What is important for purposes of this book is the *content* of the policy considerations for the particular institutional setting in which they are used for risk assessment. We will return to this topic at the end of this chapter and throughout the book.

NORMATIVE IMPLICATIONS OF THE SCIENTIFIC
UNCERTAINTIES IN RISK ASSESSMENT

A number of points emerge from the preceding discussion. First, because of substantial uncertainties, and because of the radical underdetermination of the theories by
the evidence, carcinogen risk assessment differs markedly from core areas of science. Second, in the present state of knowledge risk assessments are substantially
influenced by normative judgments. The notion of a *risk* and the *extent* of a risk
which is the outcome of present carcinogen risk assessments is at least in part a
normative notion.

Third, in addition, traditional scientific practices, which are typical of and even essential to pursuit of scientific truth for its own sake, may well paralyze risk assessment and regulatory activity. One such practice is the demand for more and better data about the substance under consideration. A particular substance may have properties and operate by biological mechanisms different from others. Yet too much emphasis on additional data can slow regulatory evaluation of the substance and can divert agency efforts from considering other chemicals. Moreover, scientists' postponing judgment until sufficient facts are available and using demanding inference standards so that they will not wrongly add to the stock of scientific knowledge will have similar effects. The justification behind such caution appears to be that by keeping the chances of scientific mistakes quite low, when one obtains a positive result, one can have considerable confidence that one's addition to scientific knowledge is not the result of random error or a mistake. In building the edifice of science, by keeping the odds of mistakes low one ensures that each brick of knowledge added to the structure is solid and well cemented to existing bricks of knowledge. Were one to take greater chances in generating new knowledge, the edifice would be much less secure. This cautious attitude is considered important in keeping the scientists from chasing chimeras and wasting time, money, and human resources.[52] (This rationale and the analogy of building an edifice may not be the whole story, for it may also be important to protect against false negatives—failing to discover an effect when it is there. Failing to discover a causal relation in the world is a bit like having a photograph with objects that were in the field of vision not recorded on the film. This seems to be just as serious a mistake about the world as a camera that adds to the photograph objects that were not in the field of vision.[53] However, it appears to receive less emphasis in science.)

In risk assessments, the cautious scientific attitude might well have the consequence that harmful responses from exposures to toxic substances are not ascertained until scientists are certain (as measured by the standards of reputable scientific inquiry) that there is such a response. In addition, scientists may be reluctant to endorse a scientific theory that would justify a particular scientific model for risk assessment inferences, unless there were sufficient scientifically defensible support for the theory. Both instantiations of this attitude could paralyze risk assessment activity, because of the many uncertainties and because the available evidence greatly underdetermines the inferences needed in risk assessment. This may not always be the case in the future, but in the present state of knowledge it is.

Fourth, common to the practices just described is an effort to prevent false positives. However, in regulation false negatives are of perhaps greater importance. The *aim* of regulation typically is to protect against toxic substances causing or contributing to human beings contracting diseases.[54] Thus, what is typically of lesser concern in purely scientific inquiry is of much greater concern in regulatory inquiries. But how much concern is justified in each case will be partly a function of the institutional setting and the aims of the activities in question.

Fifth, in any case, as I argue later, some balance of the two kinds of mistakes should be achieved for both torts and regulatory law. The proper balance is in part an institutional, legal, and moral issue depending upon the context of inquiry. We address these issues in the chapters that follow.

Sixth, the cumulative effect on regulation of many of the foregoing practices typical of research science is that risk assessment and regulation are slow. Animal bioassays are time consuming. Since the inception of the National Cancer Institute's and the National Toxicology Program's animal bioassay program, as of July 1991 there have been 367 animal bioassays (and since some of the studies are duplicative this represents fewer than 367 substances).[55] Typically it takes about two years from nomination of a substance to its acceptance for testing, two more years for the actual experiment, two years for evaluating the tumors from experimental animals, and one year for writing up the results—approximately seven years total from nomination to final write-up.[56] Five years seems to be the minimum for definitive test results from animal studies even without regulatory delay in nomination. However, this procedure is much too slow to evaluate all the suspected carcinogenic substances, for at present there are more than 50,000 chemical substances in existence,[57] some of which will be toxic, and many more are created all the time.

In addition, many new substances are introduced into commerce each year. Under procedures mandated by the Toxic Substance Control Act (TSCA) manufacturers have been required since 1976 to file a "premanufacturing notice" (PMN) when they seek to introduce a new substance into commerce. The EPA then has 90 days during which to evaluate toxicity data on these substances to see if they warrant further evaluation. From 1979 to 1987 firms filed 7356 PMN notices, or about 88 per month.[58] The rate may be as high as 150 per month.[59] Of the 7356, for about half, or 3678, firms filed notification that they were going to commence manufacturing the substances for which they had filed PMNs.[60]

Risk assessments appear to be as slow or slower than the *animal bioassays* even though they should take less time. From the late 1970s to the fall of 1990 the federal EPA and risk assessors in the state of California using similar data bases completed about 50 risk assessments of carcinogens.[61] According to the Office of Technology Assessment, as of November 1987 the EPA Carcinogen Assessment Group (CAG) had performed risk assessments on only 22 of 144 substances (15%) testing positive in at least one National Toxicology Program animal bioassay. For 61 substances for which there was even better evidence for their potency because they tested positive for carcinogenesis in three or four animal experiments EPA CAG performed only nine risk assessments (15%) between 1978 and 1986.[62] Under California's Safe Drinking Water Act of 1986 (Proposition 65), 369 carcinogens had been identified as of April 1991. However, as of fall 1991 risk assessments had been performed on only about 74 of them, leaving 295 unevaluated.[63] Thus, even if animal bioassays could be done instantaneously, risk assessments lag far behind, taking from one-half to five person years just for the detailed assessment.[64]

At present animal bioassays and the risk assessments based on them cannot possibly keep up with the introduction of new substances into the marketplace. If this is a concern, and if it points to a need to evaluate the toxicity of chemical substances more rapidly, as I think it does, then the *rate* of risk assessments becomes a relevant consideration in institutional and risk assessment design. Risk assessments should be done much more rapidly, and this would to some extent increase the number of substances subjected to regulatory evaluation. (It would not address the problem totally because the underlying animal bioassays still take about five years.) If scientists or parties to the risk assessments insist on detailed evaluation of each substance to protect against false positives, this will perpetuate the slow pace. I consider alternatives to this present practice in Chapter 4.

Seventh, a point that emerges from several of the preceding remarks is that our scientific, institutional, and policy responses to uncertainty may *promote* or *frustrate* the many institutional and social goals served by both risk assessment and regulation. It thus is important to identify appropriate institutional goals and to recognize how our responses to the uncertainties that plague risk assessment will affect the pursuit of such goals. In Chapters 2 to 4 we take up these concerns.

As we saw earlier, an attempt to make risk assessment more carefully scientific may paralyze regulation because the knowledge is not available and because almost all current regulatory laws tend to preserve the status quo until evidence for changing it is provided. (I return to this in Chapter 4.) On the other hand, if agencies expedite risk assessments and do not wait for answers to these scientific questions, then the "science" of risk assessment to some extent will rest on approximations and will be further permeated with nonscientific policies. Thus, it will be quite different from ordinary science. (Of course, even in its present state, it already differs substantially from ordinary scientific inquiry.) Scientific procedures clearly have a role in determining the health effects from human exposure to toxic substances. However, the science has to be "good enough" only for the concerns faced by the institution.

How good is good enough depends upon the many aims of the institutions, legal constraints imposed upon them, and, more broadly, matters of moral and political philosophy about the kind of world in which we want to live. The argument here is a cautionary note, a reminder, that there are substantial limitations to the extent to which risk assessments can measure up to present standards of good scientific evidence and continue to serve the aims of the regulatory institutions in which they are used. We should not confuse risk assessment for regulatory purposes with ordinary science, where the aim is pursuit of truth for its own sake, nor should we expect it always to measure up to standards of evidence required for peer-reviewed scientific journals. Instead we should recognize it for what it is and for what it can tell us, consistent with its evidentiary limitations, about the phenomena in question. In using these data, however, we should not lose sight of the many other aims of risk assessment either for regulatory activity or for purposes of tort law compensation, which should properly modify the aims of risk assessment.

A larger point is that the many uncertainties pervading carcinogen risk assessment may make it difficult, if not impossible, for scientists to remain wholly faithful to their own scientific traditions (developed in circumstances in which the pursuit of truth for its own sake was the aim) while providing data that will permit timely and justifiable regulations. More likely, fidelity to scientific tradition will produce regulatory paralysis. Thus scientists may face both cognitive dissonance if asked to participate in risk assessments and peer pressure not to modify risk assessments to serve regulatory aims. We consider this further at the end of the chapter and again in Chapter 4.

Finally, however, even if the problems mentioned did not exist, there remain more fundamental problems with aspects of carcinogen risk assessment that should be of concern to the scientists, philosophers of science, and policymakers. These are considered in the remainder of the chapter.

PROBLEMS IN THE STATISTICS OF HUMAN EPIDEMIOLOGICAL STUDIES AND ANIMAL BIOASSAYS

Epidemiological studies of human beings exposed to a substance may provide the best scientific evidence that the substance is carcinogenic at specific levels of exposure. Whether they do or not depends upon whether they suffer some possible practical and theoretical difficulties. Practical evidence-gathering problems such as poor recordkeeping, job mobility (for workplace studies), and exposure to more than one toxin may frustrate good studies. And long latency periods for diseases typically caused by carcinogens make it difficult to conduct well-done, reliable studies. However, even if *none* of these problems exist, theoretical considerations indicate that in many circumstances the design and interpretation of such statistical studies may beg some of the normative concerns at issue.

Discovering Risks

Human health risks at particular exposure levels can be detected either through cohort or case-control epidemiological studies. A cohort study compares the incidence of disease in a group exposed to a health hazard with the incidence of disease in a group representative of the general population.[65] In a case-control study, "people diagnosed as having a disease (cases) are compared with persons who do not have the disease (controls)."[66] Fewer people are needed in a case-control than in a cohort study, for only those with the disease, not those exposed to a risk factor, are the objects of examination. In either case, in a good study, a positive correlation between a risk factor and the disease indicates that those exposed will tend to develop the disease and those not exposed will tend not to develop it. Both kinds of studies can suffer from confounding, the mixing of an effect of the exposure of interest with the effect of an extraneous factor. Confounding can lead to overestimation or underestimation of the causal effect, depending upon the direction of the effect.[67] This is one of the most serious problems facing epidemiology, but because the confounding effect is not consistent, there is no one remedy. (However, in some contexts an extended search for confounders can frustrate the detection of toxins and public health protections. I return to this later.)

Case-control studies are essentially retrospective. The researcher takes a group that has contracted a disease, compares the characteristics of that group and its environment with a properly representative control group, and tries to isolate factors that might have caused the disease. Cohort studies can be retrospective or prospective. In a prospective study, a sample population exposed to a potential disease-causing factor is followed forward in time. Its disease rate is then compared with the disease rate of a group not exposed to the potential disease-causing factor. In a retrospective study the same method is employed, but historical data are used. The researcher studies the cold record of a group of people exposed to some suspected disease-causing factor over some time period to establish their disease rate. That rate is then compared with the disease rate for nonexposed groups.

Each kind of study has its advantages and its problems. Case-control studies can provide estimates of relative risk, incur little expense because the sample sizes are small, and are especially suited to the study of rare diseases.[68] They have several disadvantages. Careful diagnosis is required to ensure a properly representative control group.[69] The incidence rate cannot be derived, for there are no appropriate denominators for the populations at risk.[70] And case-control studies, like retrospective cohort studies, require historical information about their subjects, which creates problems of accuracy and documentation (discussed later). Sometimes it is difficult to separate and measure the effect of one risk factor compared with another.[71] For example, rubber workers are exposed to vinyl chloride, polychlorinated biphenyls, chloroprene, selenium compounds, benzidine and its salts, aniline, carbon tetrachloride, and benzene, all of which are either suspected or federally regulated carcinogens.[72] Case-control studies also run the risk of recall bias, since both the informant and the interviewer know the subject has the disease.[73]

In contrast, a prospective cohort study is free from recall bias. And cohort studies yield incidence rates and attributable risk as well as relative risk.[74] But cohort studies, particularly prospective ones, have their drawbacks too. They require much larger samples than case-control studies to detect the same risks and they require a long follow-up period, which increases with the latency period of a disease. Such studies are thus costly.[75] In a prospective study, subjects may drop out. In a retrospective study, they may be difficult to trace. Criteria and methods may change as the years progress. Finally, since most carcinogens have a latency period of 5 to 50 years,[76] there are ethical problems in exposing people to suspected carcinogens for the period a prospective cohort study requires.

Practical Evidence-Gathering Problems

The cost and bioethical aspects of prospective cohort studies prompt most epidemiologists to rely on case-control or retrospective cohort studies. However, there are several practical difficulties inherent in relying on historical information required for such studies. Frequently, industry data on workplace exposure to potentially harmful substances are inadequate.[77] When this is a problem, epidemiologists must resort to a worker's duration of employment as a surrogate measure of total exposure. The proper interpretation of these data, like any indirect measurement, is understandably a point of controversy, and in any event, companies often fail to keep the required information.[78] Even if such data exist, they do not necessarily reveal *which* employees actually worked in the contaminated quarters.[79]

As indicated, employees are often exposed to more than one chemical agent, which makes both case-control and retrospective cohort studies much more difficult, if not impossible, to conduct. In addition, the dosage of exposure frequently varies over time.[80]

Job mobility and population heterogeneity also pose problems. Since there is considerable job mobility in American employment, the effect of a carcinogen can easily be overlooked. Typically, the briefer the exposure, the longer the latency period of the disease, unless the exposure was at a very high dose.[81] Even if an epidemiologist has data for one population and its set of characteristics for either a cohort or case-control study, it is difficult to extrapolate to other populations and their characteristics.[82] Populations can vary in socioeconomic status, age at which exposure occurred, smoking history, and other factors that affect susceptibility and confound the studies.[83]

The long latency periods of diseases typically caused by carcinogens may be a more serious problem. Thus, even though a scientist has none of the practical problems mentioned previously and has sufficiently large samples to avoid insensitive studies (addressed later), if subjects are not followed for a long enough period, a disease effect may be missed.[84] These practical problems make it difficult, perhaps nearly impossible in some cases, to obtain scientifically respectable results to quantify health risks and to provide even the most rudimentary dose–response curve for a substance. In fact, one researcher suggested that the relevant data are missing for most chemical substances and industrial processes.[85] Thus, researchers might fail to detect a risk of concern, even when one exists, because of such practical problems.

One consequence of failing to have data about adverse health effects, even when in fact they exist, is that this *favors the legal status quo.* If employees are not protected legally from toxic substances or if they are protected less than they should be, then as long as there is no evidence of risks, even when such risks exist, workers remain unprotected. By contrast, if the legal status quo prevents the introduction of potentially toxic substances into commerce until they are proven *safe* enough for human exposure, then the legal status quo prevents commercial use of the substance. There is a problem with studies that show "safety" or no association between exposure and contraction of disease: frequently studies are too insensitive to establish the requisite claim of safety even though they do not show an "adverse effect." We pursue this further below.

Theoretical Difficulties

To illustrate the theoretical problems consider cohort observational epidemiological studies. *Observational studies,* typically relied upon to identify risks to human health from exposure to toxic substances, depend "on data derived from observations of individuals or relatively small groups of people,"[86] and in which exposure is not assigned by the investigator. They are then analyzed with "generally accepted statistical methods to determine if an association exists between a factor and a disease and, if so, the strength of the association."[87]

A wise and conscientious epidemiologist with perfect evidence, but with constrained sample sizes for studying relatively rare diseases, faces potentially controversial moral and social policy decisions in order to design and use an epidemiological study. If scientists uncritically follow scientific conventions and practices used in pursuit of knowledge for its own sake and in the requirements for publishing in reputable scientific journals, they may unwittingly have "dirty hands," contaminating their scientific results with implicit social policy outcomes and begging the policy issues at stake. In fact, the very attempt to make the science rigorous in order to publish results in reputable scientific journals may beg the regulatory questions for which the studies are done (or used). Thus, one kind of scientific objectivity leads implicitly to dirty hands. This problem is not easily avoided, for while a more sophisticated presentation of scientific results (discussed later) leaves scientists with clean hands, this merely shifts the problem to someone else.

To see these points we must review the theory of hypothesis acceptance and rejection, in order to introduce enough terminology to characterize the main risk and proof variables with which epidemiologists must work and to understand the logic of scientific proof available in this area. I focus on hypothesis rejection and acceptance because it is the traditional statistical approach used in much of science and statistics. However, epidemiology and perhaps the statistics of scientific reasoning are moving away from the model of hypothesis acceptance and rejection. Rather than explain alternatives to this model (although some are suggested toward the end of this chapter), I use it as an example to illustrate certain problems that can arise from the use of demanding standards of evidence. Thus, because the field is in flux there will be some who will not accept the model on which I focus, but I believe many (even most) still accept it and even for those who do not accept it, it can still usefully serve as an illustration of a general class of problems.[88]

In trying to determine whether a substance such as benzene is a human carcinogen, a scientist considers two hypotheses. The first (the null hypothesis, H_0) predicates that exposure to benzene *is not* associated with greater incidence of a certain disease (e.g., leukemia or aplastic anemia) than that found in a nonexposed population. The second (the alternative hypothesis H_1) indicates that exposure to benzene *is* associated with a greater incidence of such diseases.

Since epidemiology considers *samples* of both exposed and unexposed populations, by chance alone a researcher risks inferential errors from studying a sample instead of the whole population in question. A scientist runs the risk of false positives [the study shows that the null hypothesis should be rejected (and the alternative hypothesis accepted) when in fact the null hypothesis is true] or false negatives [the study shows that the null hypothesis should be accepted when in fact the null hypothesis is false (and the alternative hypothesis is true)]. A false positive is designated a *type I error,* and a false negative is called a *type II error* (summarized in Table 1–2). Figure 1–7 illustrates the relation between false positives and false negatives for two distributions of a test measure. (C is the decision cutoff such that "responses to the right of C are declared positive and those to the left declared negative.")[89] Statistical theory provides estimates of the probability of committing such errors by chance alone. The probability of a type I error is normally designated α and the probability of a type II error is designated β. Conventionally, α is set at .05 so that there is only a 1 in 20 chance of rejecting the null hypothesis when it is true.[90] The practice of setting $\alpha = .05$ I call the "95% rule," for researchers want to be 95% certain that when new knowledge is gained and the null hypothesis is rejected, it is correctly rejected.

Conventional practice also sets β between .05 and .20 when α is .05, although conventions are less rigid on this than for values of α. When β is .20, one takes 1 chance in 5 of accepting the null hypothesis as true when it is false, for example, the chance of saying benzene is not associated with leukemia when in fact it is.[91] When $\beta = .20$, the *power* $(1 - \beta)$ of one's statistical test is .80, which means scientists have an 80% chance of rejecting the null hypothesis as false when it is false.

The low value for α probably reflects a philosophy about scientific progress and

Table 1–2 False Positives and False Negatives in Epidemiology

Possible Test Results	Possibilities in the Real World of Causal Relationships	
	Null hypothesis is true: Benzene *is not* positively associated with leukemia	Null hypothesis is false: Benzene exposure *is* positively associated with leukemia
Test does not show that benzene exposure is associated with leukemia	No error	Type II error False negative β
Test shows that benzene exposure *is* associated with leukemia	Type I error False positive α	No error

may constitute part of its justification.[92] It is an instantiation of the cautious scientific attitude described earlier. When the chances of false positives are kept low, a positive result can be added to scientific knowledge with considerable knowledge that it is not the result of random chance.[93] Were one to tolerate higher risks of false positives, take greater chances of new information being false by chance alone, the edifice would be much less secure. A secure edifice of science, however, is not the only important social value at stake.

One can think of α, β (the chances of type I and type II errors, respectively) and 1-β as measures of the "risk of error" or "standards of proof." What *chance of error* is a researcher willing to take? When workers or the general public may be contracting cancer (unbeknownst to all) even though a study (with high epistemic probability) shows they are not, is a risk to their good health worth a 20% gamble?

If we think of α, β, and 1-β as *standards of proof,* how much proof do we demand of researchers and for what purposes? Must researchers be 51% sure that

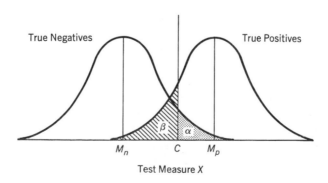

Figure 1–7 [From A. E. Ades, "Evaluating Screening Tests and Screening Programs," *Archives of Disease in Childhood* 65 (1990):793.]

benzene is a carcinogen presenting a risk to employees in the workplace before regulating it? Or, equivalently, should scientists in agencies be permitted to take a 49% chance ($\beta = .49$) that substances are not high-risk carcinogens to the populace, when in fact they might be? Such questions only precede more complex matters, for the standards of proof demanded of statistical studies have implications for the costs of doing them and for the relative risks that can be detected. The mathematics of epidemiological studies together with *small sample sizes* and *rare diseases*[94] typical of environmentally caused cancer force serious policy choices on researchers and regulators alike, when these studies are used in regulatory or tort law contexts to estimate risks to people.

In order to see some of the tradeoffs we need two other variables: N, the total number of people studied in the exposed and unexposed samples, and δ, the relative risk one would like to be able to detect.[95] Relative risk is the ratio of the incidence rate of disease for those exposed to a disease-causing substance to incidence rate among those not exposed:[96]

$$\text{Relative risk} = \frac{\text{incidence rate among exposed}}{\text{incidence rate among nonexposed}}$$

For instance, if the incidence rate of lung cancer in the nonexposed population is 7/100,000, and the incidence rate among heavy smokers is 166/100,000, the relative risk is 23.7.[97] The value of concern, δ, depends upon many factors, including the seriousness of the disease, its incidence in the general population, and how great a risk, if any, the exposed group justifiably should be expected to run. (One should note that relative risk can be misleading, if the disease rate in the general population is quite low, for example, 1/10,000,000. Thus, one needs to take into account this and other factors in evaluating the overall seriousness of the risk.)[98] With α and β fixed, the relative risk one can detect is inversely related to sample size: the smaller the risk to be detected, the larger the sample must be.

The variables α, β, δ and N are mathematically interrelated. If any of the three are known, the fourth can be determined. Typically, α is specified at the outset, although it need not be. Because the variables are interdependent, however, crucial tradeoffs may be forced by the logic of the statistical relations, as the following examples indicate (summarized in Figures 1–8 and 1–9).[99]

(1) Suppose we want to discover whether a suspected carcinogen C is associated with a particular cancer L. Suppose the incidence of L in the general population is 8/10,000, and suppose we rely upon the 95% rule. We want to be 95% sure that when no association exists between C and L, our study shows that none does. Thus we set α at .05. Suppose we also wish to have very small odds of false negatives, so we set β at .05. Thus, the chances of false positives and false negatives are equal. Suppose further that we regard a relative risk of 3 ($\delta = 3$) as a "serious" risk worth investigating for public health purposes.[100] Given these antecedent desiderata, in order to achieve them, we would have to study at least 13,495 people exposed to C, and (I assume for the sake of simplicity) an equal number who are not exposed (or 26,990 people total) to obtain statistically significant results at a relative risk of 3.[101] That very likely would be prohibitively expensive and it would be practically very difficult to follow participants. Thus, a moral consideration, the value of the most

accurate information for detecting potential harms—that is, tests with low and equal chances of type I and type II errors—can enter at the outset of a study.

(2) Next, assume everything is the same except β and sample size. Suppose we lower β to .20, so we have only 1 chance in 5 of committing a type II error. Given these values, we would have to study at least 7695 people exposed to C, and the same number who are not exposed (or 15,390 people total) to obtain statistically significant results with a power of .80 to detect a relative risk of 3. This study would also likely be prohibitively expensive and it might be difficult to find such large groups of exposed individuals. Scarcity or the impracticality of following large groups of people would still prevent the most accurate results even if type I and type II errors are not equal.

(3) Suppose we could not study the large numbers required in alternatives 1 and 2, but we could study only 2150 in the exposed and nonexposed groups. Suppose we also want to be 95% confident ($\alpha = .05$) of results favoring the null hypothesis and 80% confident ($1-\beta = .80$) of detecting an elevated relative risk should it exist, when the prevalence of the underlying disease is 8/10,000. What relative risk can we hope to detect? At best we only could detect a relative risk of 6, or two times higher than the risk we thought was "serious" enough to warrant social attention. Put differently, given the values for α, β, and N, our study could not even detect the relative risk of concern with 80% confidence, when it exists. Thus, small samples, forced by cost considerations or impracticalities and a demand for accuracy, mean that our test cannot detect the risks of concern.

Alternative 3 suggests some interesting results for "negative" or "no-effect" studies. If a study were negative or showed no effect between the chemical C and the disease L, the most that we could infer would be that the relative risk to people in the exposed group is not as high as the relative risk tested for in the study.[102] Thus, negative studies show nothing about relative risks smaller than the test can detect. Regulatory agencies regard such results as useful mainly for setting upper bounds on risks to people.

(4) The mathematical interrelations between α, β, δ, and N are flexible enough, however, to enable us to detect a lower relative risk, say $\delta = 3.8$, by making some tradeoffs. If we kept N and α constant ($N = 2,150$; $\alpha = .05$), β would have to be correspondingly raised to .49, lowering the power of the test ($1-\beta$ to .51.[103] Because $\beta = .49$, there is now, however, a 49% probability of mistaking a toxic substance for a benign substance by chance alone, when in fact the substance is toxic. The study now faces two problems. The smallest relative risk we could detect among the 2150 exposed population would be 3.8 (still slightly higher than the relative risk of concern). And we could detect that only if we were willing to take 49% odds of leaving that group exposed to a possibly harmful carcinogen. This is a morally dubious alternative, for our false negative rate is no better than the toss of a fair coin.

(5) The mathematical relations permit another alternative. Holding sample size constant, if we want to be able to detect a relative risk as low as 3.0 with 80% confidence when it exists, we could increase α instead of β. With a commitment to $\beta = .20$, the resultant α would have to be about .33 to enable us to detect a relative risk of 3.0. Now we could be only 67% confident of not incurring false positives.[104] Thus, even though we can reach statistically significant results for a relative risk of 3

by increasing α to .33, one-third of the time we run the risk of mistakenly adding to the stock of scientific knowledge. Results from such studies would not likely be published in reputable scientific journals, for the chances of type I errors are too large. Thus, we would be tolerating somewhat less scientifically accurate results in order better to be sure we could detect risks of concern.

These examples constitute the decision tree displayed in Figure 1–8. It is not immediately evident which alternative is the most attractive. Alternatives 1 and 2, although the most accurate, are excluded for reasons of cost or impracticality. Were scientists forced to adopt alternative 3 or 4, either might put those exposed to toxic substances at considerable risk because they could not detect the risk of concern (on alternative 3), or because of high false negative rates (on alternative 4). On alternative 5 scientists risk undermining the credibility of their research and increasing the risk of making a mistake in scientific research (although the odds of this are still not as high as the false negative rates in alternative 4). The logic of epidemiology, study costs, together with *small sample sizes* and *rare background disease rates* force these difficult moral choices on "scientific" research.

As striking as the preceding examples are, they only suggest the statistical problems a cohort study of a typical environmentally caused disease (e.g., benzene-induced leukemia) might pose. Alternatives 1 through 5 in Figure 1-8 assume that the prevalence of the hypothetical disease L in the general population is 8/10,000. If the background disease rate were rarer by a factor of 10, which is more *realistic* because it is the rate of leukemia,[105] then our decision tree would exhibit the even more surprising results displayed in Figure 1–9. The sample sizes required for analogues of alternatives 1 and 2 increase tenfold, the smallest relative risk that could be detected in the analogue of alternative 3 is 39, the false negative rate in the analogue of 4 greatly exceeds .60, and even increasing α to .33 in alternative 5 with $\beta = .20$ does not lower the smallest detectable relative risk below 12.[106]

There is one respect in which the preceding discussion is slightly misleading. Cohort studies require much larger samples than case-control studies to be equally sensitive. Case-control studies tend to use much smaller samples and are thus good for detecting rare diseases. The costs of using them would be much lower.[107] However, I have focused on cohort studies because the mathematics is easier to explain and the tradeoffs between sample size and relative risk are similar. In addition, cohort studies are frequently used in occupational studies, one of the concerns of this book. Case-control studies also suffer from special methodological difficulties that may preclude their use for illustrating the burden of proof problems with which I am concerned.[108]

The statistics of animal bioassays exhibit behavior similar to that of epidemiology, although the numbers are not quite as dramatic. Talbot Page has shown that if one has 50 control animals and 5 of them develop tumors at one site (e.g., the liver) while 12 of 50 treated animals develop tumors at that same site, reliance on the 95% rule would reject this as statistically significant evidence of a difference in tumor rates.[109] Nonetheless, use of Bayes's theorem and some plausible background assumptions show the tumor rate in the treated animals compared to the controls to be a matter of considerable concern.[110]

Thus, the same tradeoffs may be forced in animal studies; the rarer the disease

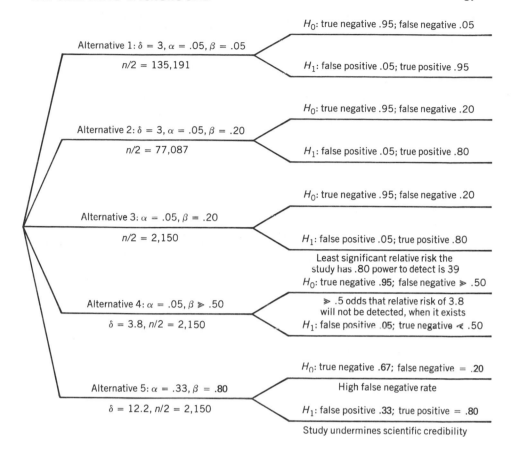

Figure 1–8 Some Choices in Conducting a Cohort Study [Numbers given for alternatives from C. Cranor, "Epidemiology and Procedural Safeguards in Workplace Health in the Aftermath of the *Benzene* Case" *Industrial Relations Law Journal* 5 (1983):372.]

rate is in control animals and the fewer treated animals in the sample with tumors, the less likely researchers would be to detect risks of concern, if they remain committed to the 95% rule. In such cases scientists should consider using higher α values to ensure that elevated disease rates in the treated groups do not go undetected because of the scientific conventions of the statistics of the studies.[111]

There is an additional problem that can arise from scientists' or risk managers' implicit commitment to certain statistical variables. Consider the effect of the α–β asymmetry in testing large numbers of substances. As long as $\alpha < \beta$ and α is in the neighborhood of .05, we are doing "better" science conventionally conceived, but we may also be protecting possibly harmful chemicals better than we are protecting human health. Suppose that we have 2400 substances to test, and that 40% of them are carcinogens and 36% of them are not, with the remainder unclear. (These percentages are similar to the results obtained from testing by the National Toxicology Program.) Now with α at .05 and β at .20, assuming there are large enough

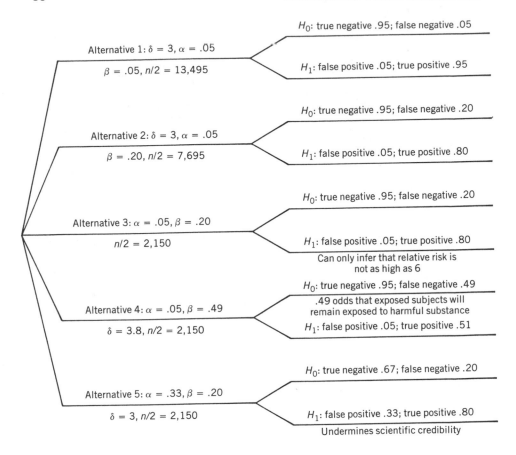

Figure 1–9 Some Choices in Conducting a Cohort Study of a Relatively Rare Disease.

study samples, epidemiological studies will result in 192 false negatives and 43 false positives. Thus, 192 substances will pose some risk of cancer to the populace (and how large a risk this is will depend upon the prevalence of the disease, the relative risk associated with the substance, its potency, and the number of people exposed). In addition, 43 substances will be wrongly regulated (or possibly banned altogether), depending upon the statutory authority in question.[112]

Moreover, although I focused on what I call the "95% rule," the burden of proof problems resulting from agency practices may be worse than I have indicated. It is probably rare that when a federal agency or an organization such as the International Agency for Research on Cancer (IARC) relies on epidemiological studies for identifying or regulating carcinogens that it relies on a single study. Several positive studies, each at the 95% confidence level, may be required before an agency is prepared to identify a substance as a carcinogen. With two such studies (assumed to be independent) at $\alpha = .05$, the chances of two such rare events occurring is .0025 $(.05^2)$. That is, the chance of two studies, each with a .05 false positive rate, both being falsely positive by chance alone is .0025.[113] To demand multiple studies the first one or two of which are positive, then, is to be exceedingly cautious. If an agency refuses to identify a substance as a carcinogen or to regulate in the meantime, this imposes costs on the potential victims while the agency gathers more information. In addition, if the agency does not address the identification or regulation of other substances, there are further opportunity costs from inaction.

Finally, although I have not discussed the effect of confounding factors in epidemiological studies, for they do not have a consistent false positive or false negative effect, there are contexts in which a continued search for them frustrates public health protections. If researchers have evidence that a substance harms human health, e.g., cigarette smoke or asbestos, but continue to search for possible confounders to explain away observed associations between exposure to the substance and contraction of disease, this can delay action and frustrate health protection. Sander Greenland notes, "One can *always* invoke unmeasured confounders to explain away observational associations. Thus, actions should not depend on the absence of such explanations, for otherwise action would never be taken."[114]

The motivation to search for confounders is similar to the motivation to require demanding standards of scientific evidence: sufficient proof to justify a scientific inference of a casual connection. Advocates of a careful search for confounders in such circumstances, such as H. J. Eysenck, seek to establish such casual connections with "proof in the sense usually accepted in science" or possibly proof "beyond a reasonable doubt" because such facts if discovered will slay "a beautiful hypothesis."[115] However, for the reasons previously discussed and for reasons to follow in the remainder of the book, such an approach poses problems in protecting public health. In my judgment Greenland's views are closer than Eysenck's to the correct approach.

Traditional Practices in Interpreting Epidemiological Studies

The preceding discussion of scientific standards of evidence indicates only some of the abstract logical tradeoffs that exist between different variables that are used in an epidemiological study. Research practices could address some of these issues.

In designing a survey, a scientist must decide which variables are to be independent and which dependent. At least one variable will be fixed: the disease rate in the general population for the disease that is the object of study. Three of the remaining four variables must be specified: sample size, α, β, and δ.

As a matter of present practice, scientists appear to specify α at some low value, typically $\alpha = .05$. The sample size is also likely to be fixed antecedently, because only a certain group of people is available for study (e.g., workers in a factory) or because costs limit the sample. If sample size is not fixed for one of these reasons, it may be chosen in light of other goals of the study which influence choice of the statistical variables, for example, α and β and the relative risk one believes is a matter of public concern. A study will then yield certain morbidity and mortality rates in both exposed and control groups, which as a matter of fact establishes an experimental to control group relative risk. Thus, in traditional practices scientists specify $\alpha = .05$, and δ is in effect fixed as an outcome of the study. Finally, scientists may use a predetermined β[116] or solve for β (or 1-β the power of the study). I assume they solve for β, for that is the more flexible and charitable interpretation of the procedure.

However, when α is set antecedently (or specified as a matter of routine) and typically at a small value ($<.05$) because the 95% rule is being used, as a matter of experimental design this creates the possibility that risks of concern may go undetected, because the power of the test may be quite low (for the reasons indicated previously).

An Alternative to Traditional Practices

Scientific practices could be different, for scientists could be more flexible in evaluating the data and consider alternatives such as 5 above, which permits both α and β to vary in value. This flexibility would permit departures from the 95% rule, in order to interpret studies to detect risks of concern. Suppose there is a study in which the fixed data consist of the disease rate (8/10,000), sample size (2,100), and mortality rate (5), compared with 1.72 expected deaths. For purposes of interpreting this information, epidemiologists could vary the values of α and β. By adopting this procedure, however, there is a problem of what the fixed data show. Any pairwise combinations of α and β in the left-hand column below will show that the study outcome is positive, for all would permit researchers to detect a relative risk of about 3. Changing the variables slightly as indicated in the right-hand column will produce a negative study.

Positive Results   *Negative Results*

α	β	δ	
.10	.49	3.0	When $\alpha <$ .10 (with β constant) or $\beta <$ .49 (with α constant)
.15	.40	3.0	When $\alpha <$ .15 (with β constant) or $\beta <$ .40 (with α constant)
.20	.30	3.1	When $\alpha <$ .20 (with β constant) or $\beta <$ .30 (with α constant)
.25	.25	3.1	When $\alpha <$ .25 (with β constant) or $\beta <$ .25 (with α constant)

Note that δ is the least significant relative risk that can be detected when it exists.

This example, together with Figures 1–8 and 1–9, shows there is considerable flexibility in interpreting the data of a study. How they are interpreted and used in certain regulatory and legal contexts will have important consequences for protecting human health.

CLEAN HANDS SCIENCE, DIRTY HANDS PUBLIC POLICY

We have seen there can be a tension between the use of the 95% (or higher) confidence rule and other public policy and moral concerns we might have. However, there is no necessity to use the cautious scientific practice—tests could be done differently. Whether epidemiologists should be committed to the 95% rule in certain contexts is a normative, or policy, question. Moral philosophers, philosophers of science, lawyers, and those in public institutions with the authority to protect our health should explicitly acknowledge and address this question. Those charged with regulating our exposure to toxic substances should consider the foregoing policy problems in the design of the study.

In addition, the reporting and use of epidemiological data given traditional scientific practices may not be as neutral and objective a project as scientists might believe. In the example just considered, what inferences one makes from the fixed data for subsequent regulatory or legal proceedings depends upon the choice of statistical variables and may have important consequences for our health and national wealth.

More important, there is not an obvious correct interpretation of the data, for the choice of values for α and β and the inferences drawn from the study *depend in part upon wider uses to which the data will be put.* But this feature of the situation commits the decision maker implicitly, if not explicitly, to making judgments that are the equivalent of moral or social policy considerations. These equivalents of moral considerations must be relied upon in order to perform and interpret the studies in question.

Alternatives are open to scientists, however, which preserve the objectivity of the science and insulate them from the "dirty" policy decisions that infect some traditional practices (when they are used uncritically).

Scientists could present their results as in Figure 1–8 or 1–9 (alternatives 3 to 5) or in a set of power function curves (Figure 1–10),[117] or with appropriate confidence intervals.[118] Recent discussions suggest there is a strong movement among epidemiologists away from tests of significance toward such presentations of data.[119] All

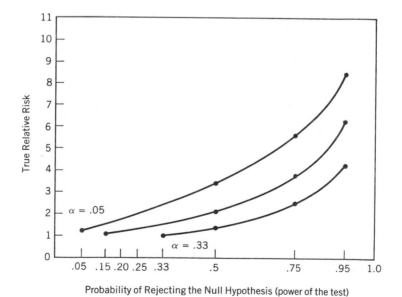

Figure 1–10 Power function curves for sample sizes 2150 and for disease rate of 8/10,000. (Numbers given for alternatives from the tables in Appendix C.)

three kinds of presentations indicate some of the *objective limits* to the data. A power function curve indicates *lower bounds* on the relative risk that could be detected (if it existed) relative to particular α values. Confidence intervals indicate the "actual magnitude of the effect as well as the precision of the estimate."[120] Common to such presentations is that they show how the data of a test place limits on what one can infer from it. In addition, the concern to display the objective limits of data serves the scientific goal of *understanding,* but this information is then used in regulatory contexts for *decision making,*[121] and those who must make the decisions must be able to understand the information and use it.

However, making inferences from data derived from small samples for rare diseases is like trying to eliminate the pucker in a wall-to-wall carpet that is too large for a room—a problem in one area can be eliminated only by creating a problem in another area.

To summarize, if scientists use α and β values uncritically in the design of their studies and the reporting of their data, they implicitly will make some of the important social policy decisions. If, on the other hand, they present the results of their studies in the most objective manner that they can, their hands remain clean, but someone else must face precisely the same social policy tradeoffs. This problem raises both public policy issues, already alluded to, and issues of professional ethics.

Professional Ethics

Who should have discretion to interpret the results of such statistical studies for regulatory purposes? Suppose that scientists have the discretion to make these decisions. There are several prima facie difficulties with this.

As we have seen, there are many concerns. While scientists use the 95% rule or confidence limits to the 95% value, they remain loyal to the conventions of their discipline and thus avoid some cognitive dissonance, but they implicitly "dirty" their hands, although they may be unaware of it, because they risk begging important regulatory issues. They similarly dirty their hands if they alter statistical variables in order to increase the power of the studies. Moreover, there are powerful peer pressures against their departing from the 95% rule. However, this last concern may be less weighty than it appears. Scientists' potential loss of credibility is a product of current practice, not necessarily a reason for continuing it. Also, the various scientific professions could develop more flexible attitudes toward scientists who participate in the regulatory process, thus removing the threat of professional criticism from them. Furthermore, even if there were the possibility of damage to a person's credibility, it is not obvious that this concern should outweigh the harm that could be done by incurring relatively high risks of false negatives in regulatory contexts and leaving the public exposed to potentially harmful substances.

A more important reason for not giving scientists the discretion to interpret studies is that this lets them determine substantial public policy questions under the guise of "science" without public scrutiny. Thus, they could implicitly dominate the regulatory process by means of the risk numbers they provide. (It is more likely, however, that scientists will use the 95% rule under the guise of "good science" to determine policy implicitly than that they will use a high α in order not to miss detecting health risks. We return to this point in Chapter 4.) This is not to impugn scientists' integrity—indeed quite the reverse: their integrity vis-à-vis the standards of their disciplines may be so great as to prevent them from seeing how a commitment to the 95% rule might implicitly skew the risk assessment process.[122] In a democracy it is arguable that such important policy issues should not be hidden in the science but settled in the somewhat more democratic forum of regulatory hearings by risk managers who are to some extent accountable to the public.[123]

These prima facie reasons suggest shifting the implicit or explicit policy decisions to the risk managers who get paid to make such highly visible and controversial decisions. Such reasons may not be overriding, however, depending upon the considerations for leaving the interpretation of studies to risk managers.

Risk managers are paid to make the *policy* choices, to take the political heat, and to protect the objectivity of scientists. (It is also likely that scientists would prefer that risk managers take such a role.) Presumably, they also have some expertise in this area. And they are in most cases indirectly responsible to a chief executive—the U.S. president or the governor of a state. Furthermore, policy considerations should enter into the regulatory debate at the risk management stage, and it is sometimes argued that there should be democratic input into the regulations. However, there are some drawbacks to leaving the interpretive discretion to risk managers.

For one thing, they may not understand the normative subtlety of risk assessment science policy decisions. Anecdotal evidence further suggests that risk managers may have considerable difficulty interpreting the most objective presentations of data, for the assessments with uncertainty bars attached or with power function curves are complicated and somewhat difficult to understand. In addition, there is some evidence that courts which review regulatory decisions may invalidate agency decisions, if the data on which agencies act present too many alternative views or exhibit too much uncertainty.[124] Thus, both reasons argue against scientists presenting risk managers and courts with complicated decision trees, tables of objective data, or data permeated with confidence intervals and giving them discretion to interpret the studies. If the failure of risk managers or courts to understand the presentation of the objective data will frustrate more effective regulation, this is a reason for not shifting the interpretive discretion from scientists to risk managers. What seems clear is that both risk assessors and risk managers should develop sensitivity to various ways of presenting data. It is even more important, however, for both to keep in mind the health-protective goals typical of most environmental health statutes so that the aims of research science do not implicitly dominate the regulation of toxic substances.[125]

This mere sketch of considerations concerning *who* should have discretion to make the implicit policy decisions in risk assessments is hardly conclusive. It is not entirely clear who should have such discretion, although I lean toward the view that policy considerations should be made explicit by the risk managers after public discussion in regulatory hearings. I consider this a bit further in Chapter 4 when discussing risk assessment in the administrative agencies. However, I believe that the public policy issues in interpreting the studies are a much clearer matter than *who* interprets the data.

Public Policy Issues

Since the reporting, interpretation, and use of epidemiological data are not normatively neutral, and we could change traditional scientific practices, we should face the use of the 95% rule in these contexts as a normative question, no matter who makes the interpretive decisions. The 95% rule in the preceding discussion serves as a surrogate or as an exemplar for the use of demanding standards of evidence for regulatory purposes. We have seen how this can adversely affect the health-protective goals of regulation. Thus, the reader should keep in mind plausible generalizations of the 95% rule as examples of how demanding standards of evidence or other scientific practices could frustrate pursuit of legally mandated goals when more sensitive and more flexible interpretations of data would better serve legal aims.

In other institutional contexts we have clearly faced such issues. In the criminal law, for example, avoiding wrongful damage to someone's reputation and well-being is so important that we spend considerable sums of money and deliberately impose difficulties on proving guilt in order to avoid wrongly inflicting harsh treatment and condemnation on the defendant. We could save money and have more unjust outcomes if we thought it worth the human costs, but we do not. Clearly a number of moral and cost considerations have influenced the institution of the criminal law. We have been quite self-conscious in debating the considerations that bear on the design and workings of the criminal law. Somewhat analogous problems arise in environmental health law concerning the interpretation and use of scientific studies; I suggest that similar debates should address these issues.

An additional problem is that the conventional evidentiary practices of science (use of the 95% rule) may well be much more demanding than the evidentiary requirements and aims of the tort and regulatory law; I discuss this issue in Chapters 2 and 4.

Thus, the problem is that the evidentiary standards of science as exemplified in the 95% rule may be much more demanding than the legal standards of evidence where the scientific evidence will be used. If this is correct, then, if regulators or courts in tort cases use conventional scientific standards, by default their science will in many cases beg the normative question at issue. Under postmarket regulatory statutes (which predominate in governing the regulation of carcinogens) a commitment to demanding scientific standards may well prevent the discovery of risks and lead to lowered protections for the public. When epidemiologists study relatively rare diseases with small samples, they will be forced to choose between high false negative rates or high false positive rates (that would be intolerable for normal scientific work).[126] If they choose to tolerate high false negatives to protect the integrity of their scientific work, they thereby favor nonregulation or less regulation by their choice of evidentiary standard.

Furthermore, given the wider aims of both the regulatory and tort law and the evidentiary standards that typically must be met in the law compared to science, there may not be good reasons in these legal contexts to require risk assessment science to meet the same evidentiary burdens as normal scientific pursuit of the truth for its own sake. Thus, I would urge that for regulatory science, agencies adopt evidentiary standards much closer to those of the legal institutions it is meant to serve (this argument is prosecuted in following chapters).

The following reasons might be offered for retaining the 95% rule even in regulatory and other legal contexts:

1. It is the prevailing tradition.
2. Scientists should be cautious about additions to scientific knowledge so that additional bricks of knowledge are well made and well cemented to the existing scientific structure.
3. The 95% rule provides a useful standard, a benchmark against which all epidemiological studies (and scientific studies more generally) can be compared. If studies were conducted with substantial departures from the 95% rule, they would no longer have a kind of automatic credibility (represented by the 95% rule) and perhaps would have to be scrutinized much more closely.
4. If the aim in an epidemiological study is to establish a causal relationship between a substance and a disease, not using the 95% rule undermines this aim.
5. The 95% rule, when used in certain regulatory contexts, protects the commercial status quo.

None of these arguments provides overriding reasons for always using the 95% rule, and several do not constitute even a prima facie reason for using it.

Reason 1, although true, begs the question whether the rule should be followed in all contexts, especially in regulatory and legal ones. The second reason, although correct for basic research that aims to add to the stock of fundamental knowledge, is less appropriate for regulatory contexts. For regulation epidemiological studies are aimed not at discovering new scientific results, but at trying to discover whether risks to health exist. And sometimes the aim is merely to confirm or deny the carcinogencity of a substance that is in a chemical class with other substances known to be carcinogenic. Neither case presents a good reason for always adhering to the 95% rule.

The third reason is an important reason of consistency, but only that. Consistency is not an overriding reason for following a certain practice, if the practice otherwise would produce bad results in a particular area. In addition, such studies are not automatically accepted at present—they receive considerable scrutiny. If the 95% rule were abandoned in some regulatory contexts this would add only marginally to the usual controversies. In particular, not using this rule would merely add to the existing complaint by some interest groups seeking to undermine risk assessments that risk assessments are not "scientific."

A related point is that a commitment to scientific standards of evidence may be the only stable reference point in debates that otherwise seem driven by political interests, policy considerations, and a good deal of uncertainty. There is much to this concern. However, it is not clear that a commitment to the 95% rule, which can beg regulatory questions, is the best way to address it. Presenting data in the most objective manner possible is quite important. But how the data are used, whether to infer a risk of concern or to infer no such risk, is clearly a matter of one's broader moral and political philosophy, or a matter to be settled in the law by the statutory guidance of regulatory law or by the procedures of the tort law. How much of a "clue" to carcinogenicity is provided by studies that fall short of the 95% rule is an important issue to be settled by the evidentiary standards appropriate to the institutional context.

Furthermore, it is a good thing to have questions of health risks decided in large part on normative policy grounds. Some degree of accuracy in estimating risks is important, but risk assessment is an inexact "science."[127] As I indicated earlier, perhaps it is much better to treat both the *kind and amount of evidence* needed to estimate a risk and the *acceptability of the risk* in part as social decisions, rather than treating the first as a purely scientific decision and the second as the only policy decision, especially when the scientific part of the decision may beg the normative issues. In addition, several researchers have found that scientists' attitudes toward their research results and toward public policy issues are substantially influenced by their place of employment. Industry scientists are more skeptical that substances pose risks of harm than are academic or government scientists.[128] Given the possibility of normative "slants" to scientists' work, it seems a better approach is to choose openly and deliberately the normative concerns we want to influence the choice of models in risk assessments and the interpretation of statistical studies. A public, community decision about these matters through the mechanism of a regulatory agency seems the appropriate approach in a democracy.

Fourth, if indeed the aim is to establish a causal relationship between a substance and a disease, then we surely want to understand this. However, whether we should wait until that is definitively established before we decide to take *action* as a matter of regulatory policy is another matter. Different kinds and amounts of *evidence* may be needed before one asserts for purposes of *understanding* the definitive existence of causal claims versus *deciding* for public health purposes what to do. For regulatory purposes one might well accept the results of epidemiological studies not based upon the 95% rule in order better to detect potentially harmful substances. For more fundamental research purposes, for example, discovering whether a whole class of substances—say, the arsenicals—appeared to be carcinogenic, one might want to have at least some of the studies established by the 95% rule.

Finally, although in fact the 95% rule used in evaluating commercially valuable chemical carcinogens may protect these commercial interests, it is for precisely this reason we should reexamine the use of the rule. For small samples and relatively rare diseases, use of the 95% rule in many circumstances may well protect commercial manufacturers and sellers of a substance better than potential victims. In weighing the balance between risks of wrongly regulating commercial substances and wrongly leaving people exposed to potentially carcinogenic substances, the latter seems the more important concern, although this depends upon the facts of the case and one's larger legal and moral philosophy. I return to these issues in Chapters 2 to 5.

In addition to the preceding rebuttals, there are some more positive reasons for modifying the 95% rule for various legal purposes. When sample sizes are small and the background rate of disease is relatively rare ($<8/10,000$), departures from the 95%

rule make it possible for epidemiologists better to detect harmful substances at a certain relative risk. This is especially important for detecting low but possibly substantial relative risks. In regulatory contexts, where a major aim of the enterprise is to predict risks to human health and to prevent them, if possible or if feasible,[129] departures from the 95% rule may better serve this preventive aim. Similarly, in toxic tort suits where the aim is to compensate victims who have probably been harmed by defendants, a departure from the 95% rule might be appropriate.

This is not to suggest that the 95% rule should be abandoned in all scientific contexts or even that it should be abandoned in all regulatory contexts. Instead scientists and policymakers should be more discriminating in their use of the rule and carefully consider the consequences of its use. In clinical trials of a drug in which the goal is to discover if a drug has therapeutic effects, the 95% rule might be relied upon, for research endeavoring to add to our fundamental knowledge about biochemical and therapeutic mechanisms should not be conducted if chances of incurring false positives are significant.[130] Similarly, when one is conducting epidemiological research to establish knowledge as a foundation for further research, one might well want to retain the 95% rule.

I suggest that on moral and legal grounds it is likely there will be reasons for departing from the 95% rule in at least these contexts:

in screening substances to try to discover those that pose harms to health;

in preventive regulatory proceedings where the major concern is the forward-looking prevention of health harm and there is little fundamental research to be gained or upon which to build[131] (pursued in Chapter 4); and

in the tort law where the typical standard of proof is not nearly as demanding as the 95% rule, perhaps courts should permit such departures (pursued in Chapter 2).[132]

For example, it might be useful for preventive health purposes for an agency like EPA or OSHA to commission a number of epidemiological studies with chances of false positives higher than .05 simply to screen for potentially harmful substances. Where there were positive results, the agency could then conduct further tests of one kind or another if additional evidence was needed, or it could randomly conduct some studies that relied upon the 95% rule in order to check for false positives.[133]

There is a generalization to the concerns raised in this section. Any specialist (at least in academic disciplines) is concerned about the validity and defensibility of his or her inferences. We tend to be cautious in drawing inferences in order to avoid mistakes. Frequently we are hyperskeptical in order to protect the field and prevent pursuit of false leads. The 95% rule is an exemplar of a minimal standard for good statistical inferences in scientific inquiries. By analogy with the arguments about epidemiology, to the extent that scientists are reluctant to conclude that suspect substances do not cause disease or death because the inferences cannot be justified on the *best* inference standards for the discipline, a debate whether to regulate or not may be begged in favor of nonregulation. By analogy with the recommendations made previously, scientists should scrutinize other scientific inferences used in risk assessment and other public policy debates to see whether regulatory outcomes are biased by evidentiary practices used in the discipline.

Similarly, the use of strict scientific inferences in regulatory contexts should be addressed as moral or social policy questions. In many cases evidentiary practices in science will beg policy questions, thus they should be examined for this *possibility*.

CONCLUSION

Several conclusions emerge from the discussions in this chapter. Sufficient uncertainties plague risk assessment procedures based on animal studies, the foundation of agency standard setting, to make risk assessment somewhat different from ordinary core areas of science. Adoption of the ideals of research science in these circumstances may result in false negatives and underregulation. Demanding standards of evidence exemplified by the 95% rule may produce a similar result. An implicit commitment to avoiding false positives may dominate risk assessment instead of some more appropriate balancing of false positives and false negatives relevant to the legal context. And, finally, assessment of the risks posed by carcinogens proceeds much too slowly to evaluate existing chemicals or to keep pace with the introduction of new substances. I will address these points in more detail in Chapter 4 when I consider the use of scientific evidence in administrative agencies.

Clearly, there will be mistakes, whether we consider estimates of risks to human health based upon animal studies or epidemiological studies. Mistakes are a result of the state of knowledge or of practical and theoretical limitations in the tools available to risk assessors. These mistakes will impose costs on someone. On whom the cost of such mistakes should fall is a normative issue. Thus, we must face the evidentiary questions posed by these different risk assessment procedures as normative matters, much as we have in designing legal procedures so that they promote and do not frustrate the larger institutional goals in which these risk projections are used. But this raises philosophical questions about the institutions involved, about what standards of evidence should be used to guide risk assessments or to establish the causal connections needed in particular institutional settings. These are topics for the chapters that follow.

2

Scientific Evidence
in the Tort Law

This chapter focuses on the use of scientific evidence in the tort law to "regulate" toxic substances. The main issue is the kind and amount of scientific evidence that needs to be offered in a tort suit in order to bring the causation issue before a jury. Some critics are recommending and some courts are requiring that plaintiffs meet quite demanding standards of evidence for this purpose.

Since the plaintiff typically bears the burden of producing enough evidence on an issue to place it before the jury, requiring that stringent standards of evidence must be met may place a burden on the plaintiff equivalent to the criminal law's "proof beyond a reasonable doubt." Such a threshold barrier for plaintiffs would substantially modify the tort law and distort the existing reasonable balance of interests between plaintiffs and defendants.

The critics' recommendations and similar court decisions seem to me mistaken, but they can be appreciated only by an understanding of the issues at stake, the kinds of evidence that can be offered, some of the philosophical goals of tort law compensation, and how the balance of interests would be upset.

We have a choice of paradigms for thinking about the use of scientific evidence in the tort law. A defensible approach is one guided not by the standards of research science, where pursuit of truth is the primary aim, but by the norms of the tort law, which strike a more desirable balance between plaintiffs' and defendants' interests than would obtain if the tort law incorporated more demanding evidentiary standards. Thus, I argue for the paradigm implicit in the tort law status quo and exemplified in a number of tort cases, and against any major changes from present practices.

In considering this topic I do not provide a wholesale critique as many have done.[1] Thus, I do not consider some of the broader criticisms brought by commentators from the political left[2] or political right,[3] although on some issues I have learned from them and in a few places I comment on their proposals. Instead the subject is a limited but crucial one for controlling carcinogens: the use of scientific evidence sufficient to establish the causation requirement for the tort law.

INSTITUTIONAL BACKGROUND

To place the issues of this chapter in a larger context we should realize that a number of legal institutions that may be used for "regulating" exposure to toxic substances. First, there is the no-institution alternative—just letting people's good judgment and the forces of the economic market guide the extent of exposure. However, this alternative is rejected by virtually everyone.[4] Second, one might try to use the contract law to permit people to contract into relationships that would provide individuals with contractual protections from toxic substances.

In a nation such as ours, which is committed "to the values of individual rights and a free society," people might normally seek morally and politically acceptable alternatives to protect our rights which "interfered least with individuals' rights and freedoms."[5] If we approach the problem from such a background it is natural to think of using the contract law to protect as many of our rights as possible, for the state could seek to protect rights by requiring that when others wish to "cross the borders defined by [our] rights,"[6] they should be required to negotiate with us for our consent to do so. Obviously this approach is particularly inappropriate to protect us from environmental harms.[7] The reason for the inadequacy is that

> the costs of negotiating with all possible victims, with all possible kinds of rights violations would be socially prohibitive and would itself lead to massive rights curtailment. . . .
>
> Thus, for these reasons we will have to protect some rights with . . . liability rules—rules requiring the payment of adequate compensation if border crossing results in harm or damage.[8]

The system of law that implements so-called liability rules is the tort law. Liberty-loving people might have a number of reasons for preferring to use the tort law to protect rights. First, even though it may be somewhat more intrusive than the contract law, for it sets public rather than privately agreed to standards of conduct, it appears to interfere only minimally with our lives. For one thing, although there are rules (developed either in the legislation or through judge-made law) that govern when compensation must be paid for the violation of rights, there are no antecedent command and control rules telling us how to conduct our lives with great specificity (as in administrative law).[9] Individuals and business firms are free to conduct their activities as they choose unless and until their actions violate existing tort law and harm others. Furthermore, once such harm occurs there is still no state intervention except to provide a forum and a set of procedures according to which legal suits will be adjudicated, settled and enforced.

Second, tort law cases will be brought only if a plaintiff victim finds the injury substantial enough to justify bringing the suit in question. If someone has been injured, that person has to make an assessment of his chances of recovering damages for the injuries suffered as well as the amount of compensation he will receive upon successful completion of a tort suit, and compare this against the time and expense of bringing the suit. The plaintiff must also compare this against bringing no suit at all. In addition, since there is no permanent prosecutorial or administrative apparatus, there are few ongoing institutional costs, and thus no community need to support an institution. Most costs are paid by the litigants, if it is in their self-interest to litigate.

Third, the tort law fits well with what we might think of as the "invisible hand conception of the law."[10] This metaphor conveys the idea that individuals in society can take care of themselves acting in accordance with their own rights and conceptions of what they need; with protections from their rights they can fulfill these needs through the goods and services provided by the economic market. And they can take care of themselves through the tort law when they are injured by seeking compensation from the wrongdoer for their injuries. The metaphor presupposes that there is no need to address and no legal problem for the law to redress, unless and until someone has brought a complaint before a court. Furthermore, when people complain about injuries, at least in the core areas of the tort law, they know the issues of and parties to the dispute. The parties to the dispute are the best ones to indicate to the court, to the legal system, what the problem is. And, typically, when a complaint is brought, someone else is charged with acting wrongly. (Of course there has been much development of strict liability rules in the tort law in the last 30 to 40 years, which has eliminated fault as a necessary condition for a successful tort suit in a number of areas of the law.[11]) In addition, in the tort law the plaintiff must establish the defendant's wrongful conduct and show the likelihood of a causal connection between defendant's conduct and plaintiff's harm. Finally, in the notion of the invisible-hand legal system, once a dispute is settled before a tort court, the issue is over. Nothing more need be done.

It is not obvious that all of the presuppositions of the invisible-hand legal system are correct, and we will consider some of them in Chapter 3. For the most part, however, I focus only on the tort law as it presently is and examine the standards of scientific evidence within it for controlling exposures to toxic torts.

There is much to be said for the tort law even if one does not fully share the assumptions of the liberals or libertarians—the liberty-loving people who might argue on political philosophical grounds for using the tort law. That is, even though it may suffer from substantial shortcomings, and even though, as I argue in Chapter 3, there are decisive reasons for not relying *solely* on the tort law to protect us from environmental harms, it has served the community well in many ways and can continue to do so in the future.

For much of our legal history it served as the main protection from environmental harms.[12] On some notable occasions recently it has served better than other institutions to protect individuals from toxic substances. In other cases it has developed in a way that improves such protections. The evidence for these claims is largely anecdotal, but it is worth sketching.

Perhaps the most dramatic illustration of the tort law "backing up" other institutions concerns asbestos litigation. Although there had been suspicions for many years that exposure to asbestos caused lung cancer and other serious lung diseases, and even quite suggestive British data in 1955, there was no definitive epidemiological evidence for adverse health affects until 1964 when Irving J. Seilikof and colleagues provided it.[13] Nonetheless, there appears to have been little regulatory action until 1972.[14] Between January and June 1972 the newly created Occupational Safety and Health Administration (OSHA) proposed and then issued regulations reducing exposure to asbestos in the workplace.[15] In 1975 OSHA issued a second notice of proposed rulemaking, but it never issued a final rule.[16] In 1984 OSHA issued a third notice of proposed rulemaking, which was made final in 1986.[17] Finally, in 1989, the Environmental Protection Agency issued regulations that required the phasing out of all uses of asbestos by the early 1990s.[18]

Thus, even though there has been regulatory action on asbestos, it has been long-delayed and some authors have argued that the tort law was really the only social institution that served well the victims of exposure to asbestos.[19] But even this protection was delayed; no tort suit was successful until 1973. That suit was brought by Clarence Borel, who began working as an industrial insulation worker in 1936.

> During his career he was employed at numerous places usually in Texas, until disabled from the disease of asbestosis in 1969. Borel's employment necessarily exposed him to heavy concentrations of asbestos generated by insulation materials. In a pretrial deposition Borel testified that at the end of the day working with insulation materials containing asbestos his clothes were usually so dusty that he could "barely pick them up without shaking them." Borel stated, "You just move them a little bit and there is going to be dust, and I blowed this dust out of my nostrils by handfuls by the end of the day. I even used Mentholatum in my nostrils to keep some of the dust from going down my throat, but it is impossible to get rid of all of it. Even your clothes just stay dusty continuously, unless you blow it off with an air hose."
>
> . . .
>
> In 1964 doctors examined Borel in connection with an insurance policy and informed him that x-rays of his lungs were cloudy. The doctor told Borel that the cause could be his occupation as an installation worker and advised him to avoid asbestos dust as much as he possibly could. On January 19, 1969, Borel was hospitalized and a lung biopsy performed. Borel's condition was diagnosed as pulmonary asbestosis. Since the disease was considered irreversible Borel was sent home. . . . [His] condition gradually worsened during the remainder of 1969. On February 11, 1970 [he] underwent surgery for the removal of his right lung. The examining doctors determined that Borel had a form of lung cancer known as mesothelioma, which had been caused by asbestos. As a result of these diseases, Borel later died before the district case reached the trial stage.[20]

Nonetheless, Borel's estate brought a suit against Fiberboard Paper Products et al. The federal district court in Texas decided for the estate, and on appeal the decision was upheld by the Fifth Circuit Court of Appeals.

This case may be the best example of the way in which the tort law has served well to provide some "protection" against environmental harm. Clearly Clarence Borel was not protected, because he died of asbestosis before his suit succeeded. However, it was the first of many successful cases against firms that manufactured and used asbestos, it called attention to the problem, it revealed an industry cover-up of the problem, and it contributed to the protection of others.[21] One commentator notes:

> The tort system was and remains the only institution capable of bringing justice to diseased workers. Were it not for the tort system, the evidence of the manufacturers' policy of silence probably would never have been disclosed, the victims of that policy would never have received anything approaching just compensation, the industry would never have been held accountable for its actions, and there never would have been any serious warning to other industries that might be tempted to ignore the human cost of their enterprise. This assessment, of course, is comparative not absolute . . . lessons about workers compensation, Congress and big business show that the tort system did not have much competition. Still it is impossible to deny the relative virtues of the tort system: it operates on the basis of private

incentives largely immune to political control and financial cooptation; it hunts for every useful scrap of material evidence; it calls upon the opinions of the most knowledgeable experts rather than relying on bureaucratic time servers; and it values human life beyond monetary losses.[22]

And the tort law can provide substantial sanctions; in 1982 the Manville Corporation, a major manufacturer of asbestos, filed for Chapter 11 bankruptcy.

The tort law has come to the fore in recent years in other cases. For example, Richard Ferebee, employed at a government research station outside of Washington, D.C., believed that he suffered lung damage from exposure to paraquat as a result of spraying the herbicide. He followed EPA-approved label instructions but believed he had contracted lung disease as a consequence nevertheless. He died from lung disease in 1982. His estate sued Chevron Chemical (the manufacturer of paraquat) for wrongful death.

Although the successful *Ferebee* suit[23] did not protect Richard Ferebee or compensate him for harms suffered, it established two important points that have made it easier for Ferebee's estate and other plaintiffs to recover damages from exposure to federally regulated toxic substances:

1. Federal pesticide regulations under the Federal Insecticide, Fungicide and Rodenticide Act do not preempt state tort law.[24]
2. The appropriate standards establishing causal claims in the tort law need not be those to which scientists would subscribe for research purposes, but need only be those of "legal sufficiency."[25]

In sum, the *Ferebee* decision explicitly authorizes state tort law to serve as a backup to federal regulatory law: it is not preempted from doing so as it would have been had the case been decided for the defendants. Thus, even if a company has observed the letter and spirit of federal administrative laws designed to protect the public (or in this case pesticide applicators) from harm, this does not prevent a victim's recovery for damages suffered, if those protective regulations are inadequate. In addition, the standards of evidence for causation are nearer those typically demanded in the regulatory law than they would have been had the court decided for the defendants on this issue. Both features of this case provide greater deterrence against other firms manufacturing and selling products which contain toxic substances that might cause similar harms and help provide a legal backup to the system of regulatory law.[26]

A third example concerns litigation about the drug diethylstilbestrol (DES). Beginning in the early 1940s several drug companies began the development of DES, a synthetic compound of the female hormone estrogen, which previously had been used as a feed additive for beef and chickens, for the purpose of preventing miscarriage. The Food and Drug Administration authorized limited marketing of DES for this purpose in 1947. The FDA stipulated that the drug contain a warning label.[27] DES has been found to cause cancerous vaginal and cervical growths in the daughters of the mothers who took the drug during pregnancy. The kind of cancer from which these daughters suffer is known as adenocarcinoma and it manifests itself after a minimum latency period of about 10 to 12 years. Further, it is a rapidly spreading, painful and deadly disease requiring radical surgery to prevent its spread.

DES also causes other precancerous vaginal and cervical growths which may spread. The treatment of such growths is cauterization, surgery or cryosurgery. Moreover, women who suffer from this condition must be monitored by painful and expensive examinations twice a year.[28]

In 1956 four doctors warned that DES, like natural sex hormones in excess amounts, might induce cancer. In 1957 Representative James Delaney (D–New York) charged in the *Congressional Record* that DES was a carcinogen, but the Food and Drug Administration denied it. By late 1959 evidence indicated that small amounts of DES induced cancer in experimental animals, including breast cancer in mice.[29] In 1971 doctors at Massachusetts General Hospital published evidence in the *New England Journal of Medicine* "linking rare vaginal cancers in young women with the use of DES by their pregnant mothers . . . 15 to 20 years earlier."[30]

Later that year the Food and Drug Administration ordered defendants (drug companies) to cease marketing and promoting DES for the purpose of preventing miscarriages and to warn physicians and the public that the drug should not be used by pregnant women because of the danger to their unborn children.[31]

The litigation brought by Judith Sindell and others did not protect them from vaginal and cervical cancer, since their suit did not precede FDA withdrawal of the drug. Nonetheless, the tort law has played some role in providing a backup to the regulatory law. FDA approval of DES did not prevent tort law recovery for injuries suffered. And court-developed rules as a surrogate for causation have made it easier for women injured by DES to recover damage awards. These rules may protect the public from similar harms caused by consumer products in the future. Such rules may provide a better deterrent to rights' violation than would have existed without this case.

The preceding evidence concerning the tort law's "protection" of victims and its backup function is mainly anecdotal, based on legal developments. There is one study founded by the Rand Corporation, however, stating that the "nine large American manufacturers generally regarded as safety leaders . . . [indicated that] products liability law administered by the courts, though generating an 'indistinct signal,' [is affecting] manufacturers' safety decisions more than do market forces or the prospect of regulation."[32] Although this study emphasizes only product liability law and does not address other tort law environmental health protections (e.g., those covering clean air and water and disposal of toxic substances), it may provide some support for increased reliance on the tort law.

Finally, there may be some cases in which the tort law awards compensation even though the firm in question is innocent—cases of mistaken responsibility ascriptions. The recent cases concerning an antinausea drug for pregnant women, Bendectin, may fall into this category, although at present the evidence is not all in.[33] And it may have been that tort law suits were wrongly decided against the manufacturers of the spermicide Ortho-Gynol, although even here the evidence is not clear.[34] Such cases sharpen the concern about appropriate evidence for causal connections.

A critical issue in all these cases has been the plaintiff's burden to show a causal connection between his exposure to a defendant's possibly toxic substance and his contraction of disease. The kind and amount of evidence needed to establish causation has become a substantial issue in recent years. If the standard of evidence is too strict, the tort law cannot well perform its protective function. If it is insufficiently demanding, some defendants will be wrongly required to compensate the victims of disease. In the remainder of this chapter I focus on this issue.

THE CHALLENGE TO PRESENT EVIDENTIARY PROCEDURES

The concept of evidence is important in both scientific and legal institutions. However, the standards of evidence in the different institutions diverge in substantial ways. For example, although scientific standards of evidence can vary substantially across different fields of study,[35] at a minimum scientists want to be at least 95% sure that they are not falsely adding to the stock of scientific knowledge when they report new discoveries or new statistical results. One of the least rigorous scientific standards of evidence, the "95% rule" appears much more demanding than most legal standards.

Although I do not yet have an adequate common conceptual framework to directly compare scientific and legal burdens of proof, the standard of adequate evidence in the tort law appears different from, and easier to satisfy than that used in scientific studies.[36] The tort law requires that the plaintiff, the party who would change the status quo, must establish his claim by a "preponderance of the evidence." This is roughly the standard that more and better evidence must favor the plaintiff's side over the defendant's side.[37] In the criminal law the moving party, the one who would change the legal status quo (typically the state), must establish its case "beyond reasonable doubt," an even more demanding evidentiary standard.[38]

The fact that different institutions in our society have different standards of evidence for their different purposes should not trouble us. However, these standards of evidence can come in conflict when scientific results are used in the law. What then should be the appropriate standard of evidence in presenting and evaluating scientific information for use in legal proceedings? Commentators differ on this subject.

One commentator, Bert Black, has launched a major attack on what he sees as the current permissive standards of scientific evidence in legal cases.[39] Black has argued

> that especially in toxic tort cases a growing number of courts now delve into the reasoning behind an expert's conclusions and require that this reasoning reflect accepted scientific practice. As society grows more tied to science and technology and more enamored of litigation this development becomes increasingly necessary. The law should seek verdicts consistent with scientific reality and with each other and it can achieve this goal only by requiring scientific evidence to conform to the standards and criteria to which scientists themselves adhere.[40]

Black suggests that a court was correct in rejecting plaintiffs' experts when they gave testimony that they "would not dare to report in a peer reviewed format."[41] And he has now bolstered his attack with a unified theory of science for the tort and criminal law.[42]

A few recent court decisions have conducted much more searching reviews of experts' scientific testimony and have granted verdicts for defendants as a consequence. Their position on expert testimony resembles Black's to some extent. In the Agent Orange cases, Judge Jack B. Weinstein argued that epidemiological studies were the "only useful studies having any bearing on causation," and that to date none favored the plaintiffs.[43] In the Bendectin cases just mentioned, four verdicts were returned for plaintiffs early on, but more recently 12 verdicts have been for defendants.[44] The latter were decided largely on grounds that the scientific evidence was inadequate despite some expert testimony to the contrary. For the most part courts deciding for defendants have relied on the absence of epidemiological results in favor of plaintiffs' cases. Some courts have ruled that scientists must be held to standards of peer review in their scientific testimony.[45] [Other courts have permitted decisions to be made without relying on epidemiological evidence, and this seems appropriate. (This will be discussed later.)]

Finally, the President's Council on Competitiveness recommended that "expert testimony be based on 'widely accepted' theories. This would eliminate testimony unsupported by scientific practice or scientific knowledge."[46] The aim is to "allow testimony on respected minority or majority theories while excluding fringe theories."[47]

The other side of the issue is the view expressed by the Circuit Court of Appeals for the District of Columbia in *Ferebee v. Chevron Chemical Company*. The justices noted that product liability law does not preclude recovery until

> a statistically significant number of people have been injured or until science has had the time and resources to complete sophisticated laboratory studies of the chemical. *In a court room the test for allowing a plaintiff to recover in a tort suit of this type is not scientific certainty but legal sufficiency: if reasonable jurors could conclude from the expert testimony that paraquat more likely than not caused Ferebee's injury, the fact that another jury might reach the opposite conclusion or that science would require more evidence before conclusively considering the causation question resolved is irrelevant.*[48]

These are two of the main sides in the controversy concerning the kind and amount of scientific evidence necessary to support legally a verdict for the plaintiff. Black urges that courts should accept only evidence that is "based upon scientifically valid reasoning, and [that] is sufficiently reliable" to satisfy legal requirements for admissibility. Courts should adhere to the standards of evidence implicit in the discipline.[49] The *Ferebee* court urges that plaintiffs in presenting scientific evidence and expert scientific testimony should be held to *legal* standards of evidence. Of course, Black's and the *Ferebee* court's standards could be the same, for the court could incorporate something akin to Black's recommendations. However, the context of the *Ferebee* opinion makes it clear that evidentiary standards to achieve tort law aims are not identical to and typically are easier to satisfy than scientific standards of validity.

Powerful forces are arrayed on both sides. On one side are plaintiffs or potential plaintiffs, public interest groups, consumer advocacy groups, all individuals who are potentially subject to such risks and who are concerned to make it somewhat easier to recover damages under personal injury law for alleged injuries suffered as a consequence of activities of others. On the other side of the same issue are defendants and potential defendants (typically corporations and industry trade associations), those who are likely to produce the risks, and, interestingly, many in the scientific community.[50] Considerable personal anecdotal evidence and reports from others suggest that scientists in order to remain faithful to the evidentiary standards of their disciplines are typically reluctant to testify to the existence of causal connections between exposure to toxic substances and harm to persons unless and until the evidentiary standards of their disciplines are met. Although in some respects this is not an inappropriate attitude, we have already in Chapter 1 seen some problems this can pose. Part of the burden of this chapter is to evaluate this claim further.[51]

Thus the issue concerning the necessary evidence to carry the burdens of proof and evidentiary demands faced by plaintiffs is of substantial moment: millions of dollars may change hands, large groups of people may be advantaged or disadvantaged, and the health of the public may also be affected. The outcome of this debate will affect procedures used to decide legal cases, either making it easier or harder for plaintiffs to recover.

LEGAL ISSUES

At issue is the notion of the "burden of proof" that must be used in a toxic tort suit for the presentation of scientific and statistical evidence. There are two distinct burdens that must be satisfied.

> The first is the burden of going forward with, or of producing, evidence. This is sometimes called the "burden of evidence" or "the production burden". . . .
> The production burden first comes into play at the very beginning of the trial. The judge and jury do not have the responsibility of investigating cases or furnishing the evidence upon which they are to be decided. Our system leaves it to the parties to do these things. If, now, neither party offers any evidence at the trial, what will happen? The answer is that one party loses. He may, therefore, be said to bear the risk of this consequence of non-production of evidence . . . he bears the burden of producing at least some evidence.[52]

The burden of production, then, is the burden of offering a minimal amount of evidence on an issue in a trial sufficient to get it before the jury.[53] Typically, the burden of production does not apply to just one "piece" of evidence, unless that is the only evidence before the court; it applies to all the evidence that bears on a particular issue.[54] The *judge* rules on whether the burden of production has been carried, and upon finding that it has not the judge may stop the trial on that issue.

Related to the burden of production is the introduction of expert testimony. In toxic tort cases expert testimony may constitute much of the evidence needed to carry the burden of production. Nonetheless, *judges* may rule independently on the adequacy of such testimony. Black in fact appears to argue strongly for a judicial evaluation of each expert's testimony independently of whether the plaintiff has carried his overall burden of production. Thus, he urges that a judge upon finding the scientific evidence or testimony inadequate might "exclude the evidence as inadmissible [to prevent the jury from considering it] or find it insufficient to sustain a verdict,"[55] thus overturning a jury verdict. He appears to emphasize careful judicial evaluation of each expert's testimony as well as the overall burden of production.

There is a second burden of proof as well.

> Where the parties of a civil action are in dispute over a material issue of fact, then that party who will lose if the trier's mind is in equipoise may be said to bear the risk that the trier will not be affirmatively persuaded or the risk of nonpersuasion upon the issue.[56]

This burden of proof is called the "risk of nonpersuasion," the "burden of nonpersuasion," or simply the "persuasion burden." Ordinarily in a jury trial the *jury* must decide if sufficient evidence has been offered to persuade the jurors of the issues at bar.

In a legal trial the burdens of proof are assigned to different parties to a dispute based upon a number of reasons.[57] However, for our purposes, we can oversimplify and say that in a legal trial both the burden of production and the burden of persuasion are typically assigned to the party who would disturb the existing legal status quo. Thus, in a tort suit this burden would be assigned to the party who complains of injury as a consequence of another person's actions and seeks compensation as a result. In a criminal trial the burden of production and the burden of persuasion are typically assigned to the state, which seeks to change the status quo by establishing that the criminal defendant has violated the criminal law and should be punished as a consequence. There are, of course, important exceptions to both these generalizations.

In the tort law the burden of persuasion is carried if the moving party establishes its case "by a preponderance of the evidence." If one could quantify the kinds and amounts of evidence that have to be offered in support of such claims, admittedly a difficult thing to do, over half of a group of judges surveyed would require that 55% of the evidence favor the plaintiff (with the rest putting it at 60–75%).[58] Others indicate that the plaintiff must establish evidence so that his claims are "more probable than not."

A plaintiff must establish each element of a tort by a preponderance of the evidence. That is, he must show (1) that the defendant violated a legal duty imposed by the tort law, (2) that the plaintiff suffered injuries compensable in the tort law, (3) that the defendant's violation of legal duty caused the plaintiff's injuries and that the defendant's violation of legal duty was the *proximate* cause of the plaintiff's injuries. The third element is the so-called cause-in-fact requirement, and it is the focus of my inquiry.

The issue with which I am concerned is *the kind and amount of evidence* that must be offered to carry the threshold *burden of production* to bring the issue of causation in a toxic tort suit before a jury and to have sufficient evidence on causation so that a judge may not overturn it. A related issue is the adequacy of expert testimony and of scientific evidence of causation in such suits. Black argues that a plaintiff's evidence is not adequate and the burden of production has not been appropriately carried until the plaintiff has established his scientific evidence in accordance with the standards that prevail in the scientific community. His stronger claim is that the evidence must measure up to the standards that prevail in peer-reviewed scientific journals. We will return to this issue later.

The issues in this debate are set by the legal standards used in different legal proceedings for the acceptance of scientific evidence. The leading rule in this regard is the "general acceptance test" for forensic science which originated in *Frye v. United States*. Defendant Frye had been convicted in the *criminal* law of a murder he claimed that he had not committed. At the trial his counsel attempted, unsuccessfully, to introduce expert testimony that the defendant had passed a systolic blood pressure detection test, a precursor of the modern polygraph lie detector. The Circuit Court of Appeals affirmed the trial court's refusal to admit this evidence and articulated its "general acceptance" rule:

> Just when a scientific principle or discovery crosses the line between the experimental and demonstrable stages is difficult to find. Somewhere in this twilight zone the evidential force of the principle must be recognized, and while courts will go a long way in admitting expert testimony deduced from a well established scientific principle or discovery, the thing from which the deduction is made must be sufficiently well established to have gained general acceptance in the particular field in which it belongs.[59]

The court's language suggests a preliminary distinction worth preserving between (1) having a general, well-supported, and recognized set of scientific principles and theories on which expert testimony is based and (2) having a certain kind and amount of evidence offered in a particular case by a particular expert witness or on a particular set of data. At a minimum the court supports the importance of having the general principle or theory well established, but it does not so clearly require exacting standards for the particular inferences made from it. This suggests that it is one thing for a court to require that the general principles of statistical reasoning be well established in the scientific community; it is quite another for a court to require that every statistical claim made in a court of law for legal purposes measure up to the evidentiary standards of good statistical reasoning for scientific purposes.[60]

Historically the courts have considered several other tests for the validity of scientific testimony in judicial proceedings. The "qualification test" used with respect to medical experts in trials "presumes, without much concern about the field of specialization, that any licensed physician is a qualified expert."[61] And, citing a 1928 authority, Black claims that doctors can testify outside their own areas of specialization. If this were a common practice, as Black points out, it could produce results that are inconsistent with scientific evidence. How serious a problem this is at present is not clear, because the practice no longer appears to be sanctioned; Black's information appears to be outdated.[62]

A third test is the "expressed certainty test." If an expert expresses sufficient certainty about the claim for which he testifies, then the court may admit it. Again, this poses a prima facie problem, for the mere fact that an expert is willing to testify that a conclusion has "reasonable medical certainty" may produce expert testimony or a trial verdict that is inconsistent with what is currently accepted in the scientific community. What is crucial here is the degree of certainty to which an expert is held in making such judgments. This expressed certainty or willingness to testify criterion was used in *Ferebee v. Chevron Chemical Co.* That court noted "if experts are willing to testify [about causation], it is enough for the jury to decide whether to credit such testimony."[63] The court did not, however, probe beneath the experts' expressed certainty that Ferebee's injuries had been caused by exposure to paraquat.

In *Wells v. Ortho Pharmaceutical Corp.*,[64] which involved a claim that birth defects were caused by a spermicidal jelly, the U.S. Court of Appeals for the 11th Circuit followed the principles of *Ferebee* and affirmed a plaintiff's verdict for about five million dollars. However, some members of the medical community have been critical of this decision. Two groups thought the court ignored "the overwhelming body of evidence" that spermicides are not teratogenic.[65] One author of a study that suggests such a connection disavows it, while three support it and a fifth author thinks it should not have been published because it was misunderstood and misused.[66] We are not in a position to judge this particular issue, but the possibility of a tort decision being inconsistent with scientific results does exist. Whether the judge's decision in *Wells* is unjust or disutilitarian depends upon wider views of the law and the role of standards of evidence within it.

Fourth, there is an additional line of cases that come up primarily in toxic tort suits in which courts have begun to examine experts' reasoning, to go behind their mere qualifications, willingness to testify, or expressed certainty about their conclusions and to "require that scientists conform to the standards and criteria of science."[67] In *Johnson v. U.S.*, a suit for damages putatively caused by low-level radiation, the court criticized the plantiffs' experts for testifying to something that "in the court's view they would not dare report in a peer reviewed format."[68] Two other courts in the Bendectin cases[69] rejected expert testimony because it failed to live up to appropriate scientific standards of the discipline. Thus, cases in some jurisdictions appear to favor more active and more searching judicial reviews of scientific evidence and some commentators, including Black, strongly encourage this trend. Call this last view the "scientific standards test," for it urges that in the case before the court experts use the evidentiary standards appropriate to the discipline in which the expert is testifying in court.

ARGUMENTS FOR THE SCIENTIFIC STANDARDS TEST

Black's view is that judges should disallow expert testimony and scientific evidence that does not meet the scientific standards test, and that judges should overturn jury decisions if the scientific evidence offered for causation does not meet the scientific standards test. He supports his conclusions with a number of arguments against the competitors and offers a general view of his own. His negative arguments, while not persuasive, point to important issues and provide some clues to an appropriate burden of proof for scientific evidence presented in toxic tort cases.

Arguments Against Competitor Views

The first argument is that some of the tests, such as "the willing testifier rule," do not require rigorous qualifications of the experts who would testify in a legal case.

As indicated previously, however, this test appears to be outdated. Furthermore, even if it is not, its shortcoming can be addressed by requiring that plaintiffs' or defendants' experts must have a field of specialization appropriate to testify about the matter at hand. (Of course, this probably occurs already since an expert's credibility would be undermined if he were not an expert in the area under litigation.) There is no need for the willing testifier rule to be modified even if it were accepted in any jurisdiction, for it can be remedied without courts' subscribing to the rigorous scientific standards of proof as Black suggests.

In addition, he claims the willing testifier rule places unwarranted trust in doctors and obscures and confuses what he sees as the real issue—the validity of the expert's reasoning.[70] Too much trust may be placed in doctors, but it is frequently difficult to judge the conclusions of an expert. However, Black's more important claim is that expert qualifications serve as a surrogate for validity of reasoning. Even if courts want to evaluate the validity of an expert's reasoning, his hypothesis is implausible—qualifications are not a surrogate for valid reasoning in an area of expertise; they are the minimum conditions of knowledge and training required for one adequately to address certain special problems. The plausibility and validity of an expert's reasoning must still be evaluated by the finder of fact—the jury or the judge in absence of a jury. Thus, even here there is a mechanism for addressing the shortcomings of the willing testifier rule without adopting Black's alternative.

Second, Black claims that the "expressed certainty test" similarly fails as a "rational approach to deciding the admissibility of medical testimony." One problem here seems to be that an expert in one trial of a case *was not* willing to subscribe to the appropriate level of certainty for his opinion, whereas in a second trial of the same case the same expert *was* willing to subscribe to the required degree of certainty (lower than in the first case) for his opinion, and thus his testimony appeared inconsistent.[71]

It is not obvious that the expert's testimony was *inconsistent*. In the first case the expert might have misunderstood the appropriate standard of evidence. Moreover, willingness to subscribe to a view with lower (75%) certainty but not with higher (95%) certainty does not make one inconsistent. That is, one can consistently subscribe to a view with substantial certainty (75%), even though one could not subscribe to it with the highest certainty (95%). In addition, it is of course better for there not to be contradictions in legal outcomes, but they are not unusual. However, where any inconsistencies result from the outcome of jury deliberations, they are not as troublesome as in other circumstances. They would be much more worrisome if they were part of the *legal rules* themselves, but they were not in this case. The expressed certainty rule merely opens up the *possibility* of inconsistent jury decisions. And there seems no good reason that evidentiary rules for scientific evidence must *guarantee* consistent jury decisions (and it is not clear they could achieve this in any case). Nonetheless, the problem raised by the expressed certainty test can be dealt with as long as judges and lawyers clearly explain to experts the difference between the degree of certainty that would be required in the scientific journals of their disciplines and the degree of certainty that would be required for legal cases in which they may be testifying. Cross-examination should also reduce inconsistencies by witnesses. In sum, inconsistencies that can arise as a result of jury decisions from use of the expressed certainty test do not require wholesale adoption of scientific standards of evidence in all legal proceedings, for these problems can be addressed in the existing adversary process.

Furthermore, Black claims that the expressed certainty standard is subject to wide variation in its use by the courts, thus "the standard has no analytic value."[72] In some courts high probability is insufficient for "reasonable certainty," whereas in other courts the mere possibility of a harmful effect is sufficient.[73]

However, even if there is variation at present in courts' interpretation of this concept, this is not necessarily a reason for rejecting the idea. If there is a consistent interpretation of "reasonable medical certainty" that will serve well, then it should be used. It cannot be condemned just because some courts do not use it carefully or properly. Many other vague concepts that serve crucial functions in the law suffer similar difficulties, but this doesn't prevent their use.

Black's third argument is that "in the toxic torts context [a court's refusal] to judge an expert's opinions according to the criterion of his or her profession can lead to results that clearly conflict with accepted scientific knowledge."[74] Thus, he notes that in *Wells v. Ortho Pharmaceutical* the court affirmed a verdict that may have ignored a well-established scientific consensus in the legal community that spermicides are not teratogenic.[75] Such a test for expert testimony should not be permitted.

There are several problems with this view. It may be that the scientific results were not as well established as Black suggests; there still appears to be some scientific disagreement about this.[76] And if they were not, then it is not clear that the decision was a bad one, if it leads to manufacturers' reexamining the constituent elements of their spermicides. Furthermore, if testimony in a particular case is contrary to well-founded scientific conclusions, it provides defendants' attorneys with a substantial basis for discrediting the experts' opinions. Moreover, the class of examples where a jury decision in a tort case is contrary to *widely* held views in the scientific community is easily distinguishable from other cases in which the outcome depends upon a "battle of bona fide experts" regarding a contested fact. Legal devices for deciding the first, such as a judgment notwithstanding the (jury) verdict in which the judge overturns a jury decision, need not be adopted in the second case, even though a judge may have some doubts about plaintiffs' experts. Finally, it is most important to note a substantial difference between cases such as *Wells* (which appear to disregard a well-established scientific consensus) on which Black concentrates and cases that may be easy to ignore in the discussion. The *Wells* case has considerable appeal because a legal decision may have been inconsistent with conclusions widely accepted in the scientific community.

However, this is, I suspect, not the typical case in toxic tort suits. In such suits there is often considerable uncertainty, dispute, and controversy concerning the factual conclusions at issue. In many cases—for example, ureaformaldahyde foam insulation, trichloroethylene, and paraquat—it may be unclear whether toxic substances cause cancer in human beings. Thus, there may be no well-established consensus in the community against which to measure the scientific validity of a particular legal judgment. Consequently, although Black may be able to score considerable argumentative points using the *Wells* case, this is probably not typical of many toxic tort cases. Where a jury verdict is inconsistent with well-established scientific theory and fact, defendants' attorneys can use this to discredit testimony or they have the use of a motion for a judgment notwithstanding the verdict.[77]

Black's final argument is that the failure to require the courts to hold experts to the standards of evidence used in science results in "the law [becoming] uncoupled from scientific reality, [producing] uncertainty and the inhibition of scientific progress."[78] He goes on to say "a manufacturer contemplating the development of new drugs or other products cannot make rational decisions if *the risk of liability is unrelated to* scientific evidence about causation."[79]

First, consider a point of clarification. How "unrelated" to scientific evidence are we to suppose legal liability is? Black appears to require that there be *no possibility* of inconsistency between scientific evidence presented at a trial and consensus in the scientific community, if there is such.[80] (Of course, there is this possibility when there is a *genuine* "battle of the experts," but in this case there is hardly scientific consensus.) But this seems too strong, for then a tort law adjudication of the scientific issues merely resembles a science court in disguise.[81] In any case, the expressed certainty rule would ensure that there was *some* relation between scientific evidence offered at trial and the state of knowledge in the discipline, but the degree of certainty would be lower than that provided by scientific certainty. Thus, there are rules intermediate between Black's proposal and the alternative he worries about. The rules endorsed by the *Ferebee* court seem to be one such set.

A second point of clarification is that the degree to which tort law rules might inhibit *scientific* progress is not obvious. They might slow *commercial* or *technological* implementation of scientific discoveries (and this can become a problem), but, whether they might frustrate scientific *understanding* or *knowledge* is less clear. Whether or not they do so depends upon *who* develops the scientific understanding—universities and research centers or private firms. Black's main concern appears to be with the technological implementation of scientific information which results in products or substances that then might harm people who use them or who are exposed to them.

Furthermore, if tort procedural rules inhibit the commercial introduction of drugs or products, this would suggest that a firm failing to introduce the product would have made the calculation that it had some nonnegligible probability of losing a tort suit signicant enough to undermine the commercial success of the product line. It is difficult to know when this becomes a problem. Armchair evidence suggests that at present we may be far from this point, for numerous new products appear to be introduced every year. Of course, if legal rules threatened commerce, a *community* could change them to encourage the development of products that firms regarded as commercially marginal if the community's need or desire for such products is great enough.[82] Furthermore, Eads and Reuter report that normally the costs of product liability cases and safety procedures in designing products are less than 1% of sales.[83] This does not seem overly burdensome.

In any case Black needs to show that technological advances *would be* inhibited when tort cause-in-fact burden of production claims *were not* based on peer reviewed scientific evidence, but that they *would not* be so inhibited when the burdens *did* have to meet such standards. That is, he has to show there is a *substantial* difference between requiring and not requiring peer-reviewed evidence, but this appears to be quite difficult to establish.

In addition, Black appears to assume that for all possible toxic tort claims, firms have tested the *safety* of their products by peer-reviewed scientific procedures. There may be some truth to this under *premarket* approval regulatory laws, such as the ones governing the introduction of drugs or direct food additives, for in these cases there is typically some kind of scientific peer review of the substance in question. In such cases, however, a *committee* of scientists typically reviews the evidence submitted by a company and makes some evaluation of the relative safety of the substance in question. It seems unlikely that at this stage of product development there will be publications *in the scientific literature* (a point Black frequently appeals to) about the relative safety of the substance. Moreover, the recent controversy over silicone breast implants suggests the possiblity that even under premarket statutes the safety of products may not be carefully scrutinized by the maufacturer or by an agency (because of grandfather clauses). Furthermore, there is a more disturbing feature of scientific practice that has just been reported. A recent study suggests that the editors of medical journals tend to favor "publishing studies that show benefits from a therapy or procedure and to reject studies that show none."[84] If this is correct, it suggests a bias that favors the introduction of new therapies. Such biases may not protect the public as well as premarket approval statutes might suggest.

When substances are subject to *postmarket* regulatory statutes, which require no premarket testing, it seems much less plausible that firms will have scientific peer reviews of their products or substances for safety.[85] Of course, out of self-protection firms would do sufficient testing to try to ensure that their products would be unlikely to harm people. But even in these cases there may be market incentives that provide contrary motivation.

However, if a firm is introducing *new* chemical substances that are subject to postmarket regulation, it is unlikely either that a committee will have peer-reviewed them or that the results will have been published in the literature. Furthermore, a firm's safety testing of substances or products is not the kind of thing that is even appropriate for peer review in scientific journals, and certainly not in the best scientific journals.

There is a further substantial drawback to Black's suggestion. If a scientist has reasonably good evidence that a substance causes harm but cannot yet justifiably publish the results in a peer-reviewed journal, judges could bar such evidence on Black's proposal. Yet, as I argue later, such evidence might be perfectly adequate for purposes of the tort law.[86] Furthermore, if a scientific *committee* constitutes adequate peer review of a substance under a premarket regulatory statute before introducing it into commerce, it is not clear why plaintiffs' and defendants' scientific experts should not testify without the support of peer-reviewed publications before a tort court, for they are testifying on the basis of their expertise much as they would in committee deliberations. Thus, Black's rule appears to work a disproportionate disadvantage on plaintiffs. It raises an evidentiary hurdle to plaintiffs by imposing rigorous scientific standards of evidence that may greatly upset the legally balanced interests in a tort suit. (See the section "A Common Conceptual Framework" later in this chapter.)

These issues point to a substantial set of interests worthy of consideration. In the toxic tort area the issues become fairly complicated, and one's intuitive "picture" of various legal institutions is important. One view of the relation might be that a company proposing to introduce into commerce a new product—for example, a pesticide such as paraquat or a consumer product such as artificial fireplace logs made of asbestos—may well have done substantial testing of the health effects of its products and may well have persuaded an appropriate regulatory agency operating under a premarket approval statute that there is no harm from such a substance *according to the appropriate standards of the regulatory statute in question*.[87] Sometimes statutory clauses are extremely demanding as are those of the Delaney Clause of the Food, Drug, and Cosmetic Act.[88] Sometimes the standards are less rigorous; for example, the standards of the Federal Insecticide, Fungicide and Rodenticide Act and of the Toxic Substances Control Act require that a substance pose no "unreasonable risk of harm" to humans or the environment.[89] However, suppose that, after this substance has been introduced into commerce, people begin to suffer various diseases and illnesses from exposure to it. How substantial must the evidence be to provide them a remedy in torts for injuries suffered?

On Black's proposal, potential victims would be left at risk without legal recourse until the scientific evidence was in and published in the literature. If a company satisfied an administrative agency that its substance was not harmful in accordance with the (probably) lower standards of evidence used in the agencies,[90] a victim claiming harm from the substance could not recover in torts until the more demanding scientific standards of tort law causation were met. Such institutional procedures would leave potential victims at risk. That is, in the words of the *Ferebee* court, the tort law would have to wait until "a statistically significant number of people have been injured or until science has had the time and resources to complete sophisticated laboratory studies of the chemical."[91] Regulatory agencies then might well permit into commerce substances which subsequently remain in commerce, free from challenge in the tort law, until, on Black's proposal, scientific evidence publishable in peer-reviewed journals was established. One legal institution would permit or even encourage products into the market, but because of the standards of evidence Black proposes, another legal institution, the tort law, would leave the public substantially unprotected and uncompensated for any injuries suffered. This possibility seems undesirable.[92]

The scenario described in the preceding two paragraphs presupposes that federal regulatory agencies have *premarket approval* or *review* authority over products. This is true for some products such as pesticides, consumer products, and direct food additives, but it is not true for many kinds of environmental toxins. Many substances are in or can enter commerce and are subject to *postmarket regulation,* that is, they can be withdrawn *only if* an agency can establish that they pose threats to human health or the environment or plaintiffs can force such substances out through suits. Under such statutes, agencies do not screen substances initially; thus, once they are in the market, any damage they might cause continues until regulatory or tort law procedures cause them to be withdrawn. In such cases Black's proposals are even less tenable.

Even under the present tort law, which does not hold scientists to Black's scientific standards test, recovery in the tort law for damages suffered may be long delayed. The history of the asbestos problem discussed at the outset of the chapter illustrates this. It is not obvious that the present tort law standards for evaluating scientific evidence, which are less demanding than those Black recommends, are the best that could be designed, but as I argue later, they are much superior to Black's proposal. Adopting Black's recommendations for the tort law would exacerbate any problems that already exist.

One's view of these evidentiary standards will depend in part upon one's philosophy of the tort law and of the relationship between the regulatory system and the tort law. If one sees the regulatory system as more protective of human health in these areas than the tort system or providing a better balance of the affected interests, then Black's argument may have some merit. On the other hand, if one sees the tort system as a partial backup for the failures of the regulatory system as some commentators suggest, then Black's view is less persuasive, because the high burdens of proof he suggests would substantially frustrate the backup function. Finally, if one sees the tort law as a kind of last-ditch insurance scheme against uncompensated harms probably caused by others, this might argue for less rigorous burdens of proof.

Consequently, major factual and institutional considerations bear on the resolution of these disputes. However, for my part I am inclined to see the tort system as providing a partial backup or court of last resort, flawed as it is, for the shortcomings of an imperfectly functioning regulatory system and thus would not want to undermine that partial protective role. And, as I have already noted, firms may be more responsive to tort law litigation than to regulatory agencies.[93] Furthermore, if one sees the regulatory system as subject to enormous lobbying pressures on the part of the companies who are the subject of the regulations or as having been in many cases "captured" by them, then this would further weaken the kind of argument that Black presents. Some of these issues are considered in somewhat more detail in Chapters 3 and 4.

Black's Proposal

Black bases his proposal for requiring more demanding standards of scientific evidence in the tort law in part on his view of the philosophy of science. He thinks that although science is not objective in some context-independent sense, scientists are still committed to striving for objectivity in science.[94] Such objectivity, including the validity of scientific views, depends upon the present "practice of science." Evidence of scientific "validity . . . hinges on acceptance" of views in the scientific community. Publication in peer reviewed journals is evidence of acceptance.[95] This "serves as an evidentiary threshold of validity. If a theory is not accepted anywhere in the literature of science, strong doubts must arise."[96] Further, he claims, judgments about the validity of the scientific research depend upon the quality of the journals involved, the pertinent scientific field, and peer review. In particular, "peer review serves at least as an initial screening process."[97]

For purposes of judging the admissibility of scientific testimony in legal cases, including toxic tort cases, Black urges that courts should actively evaluate the validity of scientific reasoning upon which testimony is based. They must determine whether the person is in the relevant scientific field, require that the experts make the reasoning clear and precise, and see that it is supported by publications in the peer reviewed literature. And, he suggests, courts should even make some inquiry into the particular journal in which results are published. Experts who cannot show that their reasoning conforms to the standards of science (presumably measured by publications in the peer-reviewed literature) "should not be allowed to testify."[98]

He strengthens this conclusion when he considers legal procedures to implement his view of scientific evidence. "When scientific evidence does not conform to accepted scientific practice, a court should either exclude the evidence as inadmissible or find it insufficient to sustain a verdict."[99] He believes that inadmissibility—prohibiting the jury from considering the evidence—is more appropriate when conclusions of an expert witness are based upon "invalid reasoning."[100] Black's specific proposal is summarized in a substitute he recommends for Rule 702 of the Federal Rules of Evidence:

> If scientific, technical, or other specialized knowledge will assist the trier of fact to understand the evidence or to determine a fact in issue, a witness qualified as an expert by knowledge, skill, experience, training, or education, may testify thereto in the form of an opinion or otherwise. When the witness offers testimony based on scientific knowledge, such testimony *shall be admitted only if* the court determines that the opinion:
> 1) is based on scientifically valid reasoning; and
> 2) is sufficiently reliable that its probative value outweighs the dangers specified in Rule 403.[101]

Notice that he would permit the judge to bar such testimony unless it was "based upon scientifically valid reasoning."[102]

It is somewhat difficult to ascertain Black's views more precisely. Sometimes he appears committed to a most extreme view that scientific testimony must be based upon publication in the best scientific journals in a field.[103] Sometimes he holds that scientific testimony must only be based upon "valid reasoning."[104] Sometimes he might be taken as holding that scientists testifying before the courts must reason as a good scientist would about the evidence. Sometimes he suggests that even publications in peer-reviewed literature are insufficient.[105]

These four accounts are subtly but dramatically different. One can reason as good scientists would, even if there are no published articles to support one's views. And scientists can fail to reason validly in publications in the best journals. In evaluating his position I focus primarily on his emphasis on testimony based on publications in peer-reviewed journals, since he returns to this frequently. From time to time I will address other interpretations his comments suggest.

Apart from Black's criticisms of other accounts of expert testimony in presenting scientific evidence and apart from contrasts between those and his, he offers few justifications for his own approach. He claims that adopting his approach will result in "improvements in fairness and efficiency."[106] He also indicates an institutional concern that his proposal will give judges better control over the outcome of trials. This is certainly true in toxic tort cases. In addition, he suggests at the outset that it is possible in any field to find an expert willing to testify on either side of a scientific issue before a court. He regards this as so undesirable that he recommends greatly increased judicial control of expert testimony to avoid it.

There is much to criticize in his proposal as well as in his suggested normative arguments for it. For one thing, his criticism of current legal theories as resting on an outdated philosophy of science, logical positivism, is not persuasive. It is not at all clear that judges believe logical positivism or, if they do (or even if it is "in the legal air"), that this contributes to the problem he identifies. And one should not think that rejection of logical positivism somehow supports his view (but I do not argue this point here).

Even if one does not endorse his philosophy of science, Black endorses various practices of science, some which should be addressed. Scientists debate theories and the evidence for them. They publish their results in journals of varying quality, and it is on these published results which courts should rely, Black claims. He appears to appeal to the sociology of science to support his view of what the law should do. This view is presently controversial in the philosophy of science and may not be the best foundation for reforming the law. One who lives by the sociology of science may also die by it. We return to some of these points later.

Second, while he emphasizes the validity of scientific reasoning, Black appears to indicate that testimony based upon publication is nearly a necessary condition of validity, but recognizes that it does not confer it.[107] Surely this is too strong. He seems so unsure about the evidentiary basis of experts' testimony that he appears willing to accept it only if it has the best public evidentiary foundation. However, scientists must be able to reason validly about scientific matters without having published their results. Every scientist who succeeds in publishing results based upon valid reasoning is in this position before the work is accepted for publication. Thus, it appears there is no reason to exclude testimony not based on publication, as he sometimes suggests.

Third, Black appears to miss, or perhaps to take advantage of, some key features of scientific practice. Scientists, and academics more generally, are typically trained as skeptics—we are taught to examine critically and very skeptically every inference argued for in journal articles and books. In many fields, the more skeptical one is, the more cautious in drawing inferences, the better one is thought to be. The extent and degree of skepticism will vary with field and even with practitioners within a discipline. This is especially true when views have not appeared in the literature and perhaps this is even more true when such views conflict with one's own scientific beliefs and commitments.[108] Frequently scientists are in the business of debunking received positions in order to establish better and more nearly correct views of how the world works. Sometimes the debunking serves the less lofty aims of providing advantages for their own positions.[109] This skepticism operates against the background of the current status quo, sometimes challenging it. When someone proposes a change in scientific understanding, scientists' skepticism resists the change until it is overcome (unless a scientist agrees with the new information or theory). Thus, a scientific view is not accepted by an individual scientist or by those in a particular discipline until some threshold of skepticism has been overcome. How substantial this threshold is depends upon the discipline and the individual involved. Furthermore, depending upon the journals, professional journal editors and referees may be even more skeptical than the average practitioner in a discipline, since they may well see themselves as the protectors of disciplinary knowledge.

If some of the historians and sociologists of science are correct, there are also less noble aims than the correct scientific view of the world that may operate, sometimes consciously and sometimes not, to make it difficult to introduce new results into the literature. Curiosity about how the world works is an important prerequisite to a scientific career. Typically, though, to be successful, scientists must receive credit for their contributions. Thus, one's individual skepticism may resist competitors' views more than one's own. Self-interest may sometimes motivate skepticism. Moreover, scientists must make their findings public, but they must be cautious enough in "going public" that they are not seen as irresponsible.[110] Mistakenly adding to the knowledge of the world both harms a scientist's own career and "harms everyone who uses [the] faulty contributions."[111] Thus, there are powerful motivations of self-interest and protection of the scientific enterprise to guard against mistakenly adding to the present stock of scientific knowledge. These forces may militate against research scientists arguing for bold theses.[112] What is much more mundane and more relevant to the issue at hand is that these forces may also work against scientists providing early warning about possible harms from exposure to possibly toxic substances, if these are not part of the received view. They must come forward with their views, and to some extent risk their standing in their peer group in order to call attention to such harms.

Black's proposal uses the skepticism of scientific disciplines and these less visible but powerful sociological forces to protect the status quo of scientific knowledge for legal purposes. If a plaintiff believed that he had been injured by a substance, say, asbestos, at a certain time, say, in 1955—after the British had substantial clues to the harm from asbestos exposure but before a good epidemiological study had been completed—then scientific skepticism works for the defendant. If there has been no *publishable* evidence to date that a substance harms humans, then according to Black a tort suit should not be decided for a plaintiff who brings suit that the substance harmed him. This suggests that only when there is a substantial consensus as to harm among the appropriate experts is a finding for the plaintiff warranted. Thus, until the skepticism of the scientific community, as well as the skepticism of journal editors and referees, is overcome, plaintiffs should remain uncompensated. Black either fails to understand this point or chooses to use it for defendants in toxic tort and other litigation. To overstate this point somewhat we might say that, at least in his most extreme moments, Black leaves the *legal sufficiency* of evidence relevant to toxic tort litigation to skeptical journal editors, skeptical scientists, and possibly even to the skepticism of scientists opposed to a finding of harm.

Fourth, apart from his views about science and his obvious reliance on scientific skepticism, Black's position raises potentially controversial issues of legal philosophy. For one thing, in several places he argues that *courts* (read "judges") should control the legal sufficiency of scientific testimony, based upon the *validity* of scientific reasoning. He urges that judges be more activist and interventionist in screening the testimony that the jury considers (by granting more motions *in limine*[113] as well as more motions to dismiss and motions for directed verdicts) and in controlling the verdicts they reach (by granting more judgments notwithstanding the verdicts).

This appears to be a recommendation that judges do much of the work for the defense. Instead of the defense presenting evidence *to the jury* that the plaintiff's expert is mistaken, reasoning badly, or testifying contrary to a well-established body of scientific evidence and letting the jury decide the issue, he proposes taking some of the decisions from the jury and having them controlled more closely by judges. This raises much larger philosophical issues about the proper roles of judge and jury, which cannot be developed here, but consider a few points.

Black appears not to trust juries to make the correct decision, or not to be correct enough of the time, to leave such judgments to them. The acceptable correct decision rate for juries in tort cases is not clear, but his proposal would substitute judicial control and scientific skepticism for jury decision in these areas. This may well strongly favor defendants.

There might even be an increase in trial length, for it appears on Black's favored approach that a judge would have to screen scientific testimony before a jury hears it, then it would be heard a second time by the jury. This hardly seems efficient—one of the virtues he claims for his proposal.[114]

Further, while judges can and certainly have made judgments about scientific acceptance of scientific theories and facts, it is not clear juries cannot make similar judgments when presented with much the same information. Assigning this job to judges increases their work and places greater demands on their out-of-court evaluation of a case. The more active a judge is in screening a plaintiff's evidence, the greater the importance of the *judge*'s attitudes, beliefs, and opinions become as opposed to those of the jury. This is important, for unlike a jury, a judge cannot be examined for bias and attitudes. Judges are assigned to cases; juries are chosen by the litigants.

Moreover, the allocation of responsibility between judge and jury to decide these issues raises a broader political philosophical point about democracy. Juries tend to represent the community's concerns, as the democratic appliers of the law, whereas judges tend to be the government's representative as providers of "orderly supervision of public affairs."[115] Judicial power to exclude juries from considering the factual issues of toxic exposure deprives the more democratic body from participating in that aspect of the decision. At a time when there is increasing public concern about exposure to toxic substances, increasing the burden of production which the judge controls seems ill-timed and wrongheaded. And, given the normative aspects of assessing and controlling toxic risks, public input via jury decisions seems quite important.

This review does not exhaust evaluation of Black's proposal. However, together with the earlier evaluation of his criticisms of others, it does raise substantial doubts about the plausibility of his recommendation. It also raises more fundamental issues about the relation between the law and the role of scientific evidence in it. I turn now to one of these issues.

A COMMON CONCEPTUAL FRAMEWORK FOR EVALUATING
SCIENTIFIC AND LEGAL BURDENS OF PROOF

In Chapter 1 we saw several ways in which scientific practices might frustrate legal goals in the regulation of toxic substances. Scientists unwisely demanding more and better data, withholding judgment until there is sufficient research or until all possible confounding factors are ruled out, and using demanding standards of evidence can all prevent the discovery of potential health harms. In the tort law these, taken separately or together, if rigidly adhered to could delay for years tort law recovery for harm suffered from toxic substances.

In the use of *statistical studies* exemplified by epidemiology, there can be a tension between adhering to the scientific standards used for pursuit of scientific truth for its own sake (and which is required for peer reviewed journals) and pursuing the policy or social objectives for which the scientific results might be used in legal proceedings. *Scientific evidence* need not in every case exhibit all the properties of epidemiological studies. The *special* property of such statistical studies is the false positive/false negative error tradeoff that is forced by mathematical equations. There may be other scientific evidence where such tradeoffs do not exist. In such cases the *special* problems discussed earlier may not exist, but the difficulties considered later of adhering to scientific standards of evidence (as suggested by the 95% rule) in legal cases remain. Thus, although this tension *need* not result, in many toxic tort cases it will. The epidemiology example both exemplifies this and represents the more general problem that demanding standards could cause.

In toxic tort litigation there typically is not one "piece" of evidence that is determinative of an issue, as the epidemiological example suggests. Many pieces of evidence may bear on whether a plaintiff has carried his burden of production. However, since one major issue is the admissibility of individual pieces of evidence or expert testimony, the example is somewhat more representative. In the discussion that follows I use this example as illustrative of the more general problem that demanding scientific practices can cause.

To better understand the burden of proof issues and to evaluate various recommendations, we need a common conceptual framework to compare the different standards of evidence. The notions of probability central to the scientific examples and to the legal burdens of proof—the burden of production and the burden of persuasion in tort cases and the "beyond a reasonable doubt" burden of persuasion in criminal cases—are not identical, thus we must have a common framework with which to compare them.

Recall that the probabilities of false positives and false negatives are *conditional probabilities*. A false positive (FP) is the conditional probability that by random chance alone the results of a statistical test will show that the null hypothesis should be rejected when in fact it is true. A false negative (FN) is the conditional probability that by random chance alone the results of a statistical test will show that the null hypothesis should be accepted when in fact the null hypothesis is false. Expressing these symbolically, a false positive is $p(-H_o/H_o$ is true), and a false negative is $p(H_o/H_o$ is false).

The preponderance of evidence test in toxic tort cases, however, is a relation between the kind and amount of evidence the plaintiff offers and the kind and amount of evidence the defendant offers, not conditional probabilities. According to *McCormick's Handbook of the Law of Evidence,* the moving party, typically the plaintiff, must carry the *burden of persuasion* so the jury finds "that the existence of the contested fact is more probable than its nonexistence."[116] In short, the moving party must show that the conclusion he wishes to draw from his evidence has a (slightly) greater probability of being true than the conclusion for which his opponent argues. If it has a greater probability, the moving party wins the contested issue; if it does not, the opponent wins.[117] An obvious candidate for comparing the conditional probabilities of FPs and FNs in science with the legal burdens of proof is a framework based upon the relation between FPs and FNs. These notions in effect express possible error rates of the procedures in question. In the scientific case the error rates are dictated by sample sizes and the relative risks one desires to detect. In the legal realm the error rates are suggested by the norms of the law.

Table 2–1, summarizes the qualitative idea of false positives and false negatives for a tort case. We can think of the null hypothesis as the claim that the defendant in a tort (or criminal) case is not responsible under the applicable legal rules, since in a tort case the plaintiff or complaining party is the one who seeks to disturb the status quo ante (in a criminal case it is the state that disturbs the status quo) and is the one who typically has the burden of production (going forward with evidence) and the burden of persuasion. In the scientific analogue researchers accept the status quo and any departures from it must be justified by the presentation of evidence. The alternative hypothesis in the legal case is the claim that the defendant should be held accountable; in the tort law this is the claim that the defendant should be made to pay compensation to the plaintiff, while in the criminal law this amounts to the claim that the defendant is guilty of violating the criminal law.

In the real world, there are two possibilities: the defendant either should be held accountable or not under the applicable law. And there are two possible outcomes from a legal trial: either the defendant is held accountable or not. However, just as in the scientific case, the "institutional test"—the legal trial—does not necessarily show what is truly the case in the real world. Mistakes are possible.

Although Table 2–1 is analogous to Table 1–2, there are no equations to generate the probabilities of false positives and false negatives for trials. We can find a basis of comparison between scientific and legal burdens of proof, however, by considering some of the error norms of the law and comparing them with some plausible error norms of science.[118]

Begin with the criminal law. In the criminal law it is thought that it is better for ten guilty men to go free rather than for one innocent man to be wrongly convicted. This is typically taken to express the idea that the probability of false negatives to the probability of false positives in the criminal law should be in an approximate ratio of 10:1 *given the magnitude of harm done by convicting an innocent man versus the magnitude of harm done by letting guilty parties go free.*[119] This seems to express the idea that, given the normative interests at stake in wrongly convicting an inno-

Table 2–1 False Positives and False Negatives in a Tort Law Trial

	Possible Correct Tort Law Relationships Between Adversaries	
Possible Trial Outcomes	Null hypothesis is actually true: the defendant is not responsible under applicable law	Null hypothesis is false, alternative hypothesis is true: the defendant is responsible under applicable law
Null hypothesis is accepted; defendant is found not accountable	No error	False negative
Null hypothesis is rejected (and alternative hypothesis is accepted); defendant is found accountable	False positive	No error

cent person compared with wrongly letting a guilty person go free, to balance the scales of justice between prosecution and defense, the expected utility of a false negative must equal the expected utility of a false positive. The general relationship is

probability of a false negative $\times$ magnitude of a false negative $=$ probability of a false positive $\times$ magnitude of a false positive (1)

To offset the great loss that would be suffered by a defendant from a mistaken conviction, the criminal law has developed compensating norms for the burden of proof which must be met to convict a defendant. It should be approximately ten times as hard for the prosecution mistakenly to convict as it is for the defendant mistakenly to go free, given the relative costs of wrongful conviction versus wrongful exoneration. We might think of the burden of proof that the prosecution faces (it must establish its case "beyond a reasonable doubt") as a procedural rule designed to secure *in an approximate way* and *other things being equal* the false positive/false negative error rate. Whether it will successfully do so depends upon the kind and quality of evidence that is offered by both sides in a particular case and by the impartiality of the jurors and judges involved. Thus, whether the procedural rules are successful in achieving the normative error rates depends upon a number of contingent circumstances.

Contrast the criminal law with the tort law. Wrongfully holding the defendant responsible and making him pay damages is not thought to be such a serious wrong as a mistaken conviction in the criminal law. In fact the normative interests at stake between plaintiff and defendant are thought to be approximately equal (or possibly reversed to some extent[120]). Mistakenly holding an innocent defendant accountable is not thought to be much worse than mistakenly failing to compensate a deserving plaintiff. To set the scales of justice approximately equal between plaintiff and defendant, the general formulation (1) again seems appropriate. Now, however, the chance of wrongly holding a defendant accountable is approximately equal to the chance of wrongly failing to award compensation to a plaintiff.[121] Thus, we can simplify the normative relationship between false positives and false negatives in the tort law and say that the probability of a false positive is (approximately) equal to the probability of a false negative in the tort law. Symbolically we can represent this as

$$P_{FNL} = P_{FPL} \qquad (2)$$

where P_{FNL} is the probability of a false negative, and P_{FPL} is the chance of a false positive in the tort law. The burden of proof rule for torts, that the plaintiff's case must be established by a "preponderance of the evidence," slightly more than 50%, may be thought of as a way of approximating the tort law error rates. Again, whether this procedural rule succeeds in producing the normative error rates is a contingent matter.

When we turn to science the issue is somewhat more complicated, but the norms regarding the ratio of false positives to false negatives appear to be similar to those in the law. Chapter 1 emphasized that scientists seemed more concerned to prevent false positives than false negatives. This apparent concern with false positives might mislead us slightly as to the *norms* of science. In hypothesis testing we saw in Chapter 1 that the *most accurate studies* are those where the false positive and false negative rates are *equal* and *low*, for example, $\alpha = \beta = .05$. Such studies have both high statistical significance and sensitivity. These seem proper norms, because scientists should want both to avoid mistakenly adding to scientific knowledge and to avoid missing evidence of causal connections when it is there.[122] Each seems important.

However, in many circumstances concessions to the costs of experiments or to the sizes of samples available for study might force one to choose between false positives and false negatives. When practical circumstances force a retreat from the ideal, it is quite reasonable for scientists to choose to protect more against false positives than against false negatives, for the reasons already considered. Nonetheless, *as a norm* it seems that science aspires to equal protection against false positives and false negatives. We can express this symbolically as

$$P_{FNS} = P_{FPS} \qquad (3)$$

where P_{FNS} is the chance of a false negative and P_{FPS} is a chance of a false positive in science.

We can now use equations (2) and (3) to compare the error rates of science and the law and to see the effect that one would have on the other. Dividing equation (2) by equation (3) we can produce (4), which is the ratio of the probabilities:

$$\frac{P_{FNL}}{P_{FNS}} = \frac{P_{FPL}}{P_{FPS}} \qquad (4)$$

What (4) says is that the ratio of the probabilities of false negatives in the law to false negatives in science is identical to the ratio of the probabilities of false positives in the law to false positives in science. Reorganizing (4) we see that the ratio of the probabilities of false negatives to false positives in the law is equal to the ratio of the probabilities of false negatives to false positives in science:

$$\frac{P_{FNL}}{P_{FPL}} = \frac{P_{FNS}}{P_{FPS}} \tag{5}$$

This formula turns out to be quite revelatory, as particular examples show. Intuitively, what (5) says is that if we want the probabilities of false positives and false negatives to be (approximately) equal in the law and the probabilities of false positives and false negatives to be equal in scientific experiments for normative purposes, then the way to achieve this is to set the ratio of the probabilities of false negatives and false positives in the law equal to the ratio of the probabilities of false negatives and false positives in science. We can then use the ratio of the probabilities from one institutional setting to compute the analogous ratio of probabilities necessary in the other institutional setting to achieve these goals. This relationship permits in an abstract way a comparison of the ratio of error rates of one institution with the ratio of error rates of another.

A somewhat simpler argument leads to the same conclusion. We do not need ideal error rate ratios for science to see the influence of scientific outcomes on the error rate ratios of the law.[123] As long as the scientific error rates influence legal error rate ratios, all we need for purposes of the position argued for here is the relationship summarized in equation (2).[124]

To see the distorting influence scientific standards of evidence can have on the law, consider an epidemiological study performed under less than ideal conditions. Suppose the false negative rate in an epidemiological study is .5 (equivalent to a coin toss) and the false positive rate is .05 (a typical value). These numbers are similar to alternative 4 in Figure 1-8. Sample size, background disease rate, and the relative risk one wants to be able to detect impose certain error rates on such a study (assuming a scientific commitment to a type I error rate equal to .05). Now suppose, contrary to what is typical in most legal cases, that the outcome of the legal case would depend only (or at least substantially) upon this one piece of scientific evidence for causation in a tort suit. What does this imply for the error rates of legal procedures if the courts should adhere to the demanding standards of scientific evidence ($\alpha = .05$), as Black recommends? We can see this by substituting these numbers into formula (5).[125] The resulting ratio of the probabilities of false negatives to false positives in the law would be 10 to 1, summarized as follows:

$$\frac{P_{FNL}}{P_{FPL}} = \frac{.50}{.05} = \frac{10}{1} = \tag{6}$$

where

$$P_{FNS} = .50 \text{ and}$$
$$P_{FPS} = .05$$

Thus, if we use the false negative/false positive ratio from a realistic statistical example in science to dictate the false negative/false positive rates in the law, we get a ratio of 10:1. How do we interpret this 10:1 ratio? The aphorism from the criminal law that it is better for ten guilty men to go free rather than for one innocent man to be wrongly convicted expresses one view of the ratio of the probability of false negatives to the probability of false positives in the criminal law; it is identical to the outcome in (6).

The consequence is that if an epidemiological study were to be conducted in accordance with a typical scenario (but with fairly high false negative rates) and the probabilities from that used in the resulting legal process to establish the ratio of the probability of false negatives to false positives in the law, this would produce an FN/FP ratio identical to that typically required in the criminal law. Another way of putting this point is that the probability of a false negative in the tort law would thereby be ten times greater than the probability of a false positive. Whereas the 10:1 ratio seems appropriate for the criminal law (a false negative means that a guilty person goes free), it seems much too demanding for tort suits.

If the false positive rate in science is lower than .05 (e.g., .01), as it often is, the FN/FP ratio in the law would be even greater and thus more seriously unsuitable for either the burden of production or the burden of persuasion in the tort law. If the false negative rate in science were lower than .50 (e.g., .20, the typical false negative rate used for many examples in Chapter 1), the FN/FP ratio in the law would be lower (4:1). Such a ratio still substantially disadvantages the plaintiff compared with a defendant, because of the threshold burden he faces. Sometimes the false negative rate is even higher than .50.[126] In such cases this means there are even greater odds by chance alone of failing to detect an association between exposure and disease when it exists. This would distort tort law error rates even further from the norm.

The upshot of these examples is not just that the standards of evidence in science and in the law are substantially different. The examples also try to establish in a rough quantitative and symbolic way how much the evidentiary standards differ. If scientific standards were to dictate legal standards for the burden of production as Black recommends, this would distort legal relationships.

Two qualifications of the preceding argument are needed. The first is that the hypothetical case just considered may appear a bit unrealistic and to that extent somewhat misleading. Typically, a plaintiff would try to establish the burden of production by more than one scientific study or by the testimony of more than one expert. To carry this burden the plaintiff must provide enough evidence of the appropriate quantity and quality to satisfy a judge that this burden has been carried. The plaintiff must show that all the evidence taken together is sufficient for this purpose. Plaintiff's case is not likely to rest on a single piece of evidence.

The hypothetical example is not as misleading as it might appear, however, for several reasons. For one thing, Black's recommendations are addressed to the admissibility of "pieces of evidence" or the testimony of particular experts. Thus, he would exclude expert testimony or a piece of scientific evidence if it failed to measure up to peer review standards. Insofar as his proposal is directed against standards for the admission of single pieces of evidence or the testimony of a single expert, the hypothetical example developed here is apt.

Second, although the plaintiff may present more than one piece of evidence, only one or two of these may be critical to a decision. When this is true, the hypothetical study is more nearly appropriate for characterizing the issues. For many tort cases the outcomes seem to depend on a small number of studies and frequently a small number of expert witnesses on the issue of causation.

The hypothetical example just considered and examples discussed in Chapter 1 should suggest substantial caution about the use of negative studies in toxic tort cases. A legal requirement that scientists base their testimony on typical scientific standards of evidence could not only distort the tort law relationships between plaintiff and defendant, but in many circumstances it would almost certainly result in a study much too insensitive to detect even high relative risks that might be present. If the true relative risk from a carcinogen similar to benzene were just greater than 2, so that tort law compensation would be justified if it could be detected, a cohort study big enough to detect such a rate would have to sample several hundred thousand people exposed to the substance, and it could still have a fairly high false negative rate. This is another hidden consequence of Black's proposal.[127] Even a case-control study which requires smaller sample sizes might have to be relatively large to detect such small relative risks.

A recent toxic tort case illustrates the problem with negative epidemiological studies. A number of people were exposed to hazardous substances leaking from a landfill operated by Velsicol Chemical Corporation. The plaintiffs contracted a number of diseases, including cancer, and sued Velsicol for compensation. Velsicol produced an epidemiological study taken from workplace exposures which showed there was no increased risk of kidney cancer from the exposure. The plaintiffs' expert, a distinguished scientist retired from the National Cancer Institute, carefully showed that the study because of its small sample had a probability of only .303 of detecting a twofold increase in kidney cancers and a .42 probability of discovering a doubling of bladder cancers. The district court holding that such a study had "no weight" in that case had its major holdings upheld on appeal.[128]

There is an additional qualification to the argument comparing scientific and tort law burdens of proof. I have compared a typical scientific burden of proof with the tort law burden of persuasion. However, typically the important issue in toxic tort suits is the plaintiff's burden of producing evidence, or the burden of production. The stringency of the burden of production need not and may not be identical to the stringent burden of persuasion. *McCormick on Evidence* indicates two main tests courts have utilized of whether the burden of production is carried: a jury *can conclude* that a preponderance of the produced evidence favors the party with the burden or a jury *cannot help but conclude* a preponderance of the produced evidence favors the party with the burden.[129] Both are the same as or less stringent than the "preponderance of the evidence" burden of persuasion. Thus, any assumptions I have made about the (approximately) equal burdens of proof between plaintiffs and defendants seem reasonable. And, although a few courts appear to have argued that for circumstantial evidence, the evidence produced must be "so conclusive as to exclude any other inference inconsistent with" it, *McCormick* argues it is "misplaced in civil litigation."[130] Apart from a very few cases mentioned by *McCormick* advocating this view, there appears to be no support for the idea that the tort law burden of production should be the same as the criminal law's "beyond a reasonable doubt" to which Black seems committed.

There is a final point about error rates. My intuition is that the false positive rate in science is substantially lower than the false positive rate in the law in some absolute numerical sense. If this is correct, it provides further argument for the distorting effects of scientific standards in the tort law. I do not have a definitive argument, but some considerations seem appropriate. For one thing, the law, unlike science, is not obviously concerned to add cautiously to a stock of knowledge through legal decisions at the trial court level. Thus, there is no need for a legal commitment to a false positive rate as low as .05. It is important that plaintiffs and defendants overall be treated roughly *equally,* but the absolute false positive and false negative rates need not be set at a low predetermined level as in scientific inquiry. And there seems to be no good reason to recommend such changes in the law, as I have already argued. In addition, the difference between the burden of proof rule in torts and the burden of proof rule in the criminal law suggests that there is a lesser concern to preclude errors in the tort than in the criminal law. I cannot find a numerical value for this difference, but it seems clear.

The preceding considerations indicate that the error rates of the law and science are substantially different. If we adopt plausible numbers for false negatives and false positives from the scientific tradition, the resulting ratio of probabilities that would be produced for the burden of production for scientific evidence much more closely resembles the burden of persuasion required in the criminal law than the burdens typically required in the tort law. Thus, Black appears to support a *burden of production* in toxic torts that is as high or almost as high as the burden of persuasion in the criminal law. This problem is further exaggerated if his proposal applies not to the overall burden of production, but to each instance of scientific testimony. We can make this point in a dramatic way by noting that *evidence of causation could not be admitted to go to the jury unless it was sufficient (or nearly sufficient) to satisfy the equivalent of the demanding persuasion burden of the criminal law.* This is surely a substantial and unfair barrier to plaintiffs, and it introduces substantial distortions into the tort law.

The idea is that plaintiffs would have to overcome threshold burden of production barriers equivalent to the criminal law's proof beyond a reasonable doubt in order to place the evidence before a jury. If the plaintiff cannot carry this burden, the plaintiff loses. In such circumstances plaintiffs would find it difficult to "get their cases going" and to receive a hearing before a jury of their peers.

Furthermore, as we have seen, in science the overriding concern is to avoid false positives, whereas the law attempts to strike some appropriate balance between legal false positives and false negatives. Importing the scientific standard into the law distorts the latter balance of interests.

Finally, given the foregoing arguments, there appears no good reason to require that scientific testimony in the law satisfy the requirements of peer reviewed presentations in science, for there are obvious alternatives to address Black's concerns.

AN ALTERNATIVE VIEW

There may be a temptation to argue that if one does not have scientific or statistical evidence that protects 95% against false positives, then one does not have good evidence. This is a mistake, for it confuses the requirements of evidence for one area of human inquiry with the evidentiary requirements for another area of human inquiry which has quite different purposes. Thus, the preceding considerations suggest the conclusion of the *Ferebee* court: "In a courtroom the test for allowing a plaintiff to recover in a tort suit . . . is not scientific certainty but legal sufficiency."[131] Something like the traditional legal requirement of "proof to a reasonable degree of medical certainty" seems the appropriate standard for the burden of production in toxic tort suits. The plaintiff should have some threshold burden of production "to prevent verdicts based upon speculation and conjecture,"[132] and to help preclude frivolous claims, or "junk science," but not the unreasonably high burden Black suggests.[133]

Several reasons argue for not departing from the *Ferebee* status quo and for not adopting a more stringent standard. Black's arguments against present practices are not persuasive. And his proposal would distort tort law relations. Moreover, the *Ferebee* court's rule is more sensitive to the institutional context and how that should guide our legal epistemology.

The *Ferebee* approach is more consistent with traditional aims of the tort law. These rules make it somewhat easier for plaintiffs to recover than will Black's approach, thus serving the compensatory and deterrence aims of the tort law. They will also likely result in greater safety from potentially toxic substances. However, there is a possible downside to the *Ferebee* status quo: as Peter Huber argues, it may result in new products and technologies being regulated too stringently compared to older, even more dangerous technologies, and it may inhibit the development of new products.[134]

I have already commented on the last point. Whether the *Ferebee* view compared with Black's view will decrease our safety from toxic substances has been addressed indirectly by others. In an excellent discussion, Gillette and Krier, addressing Huber's safety concerns, argue that while *procedural rules* in the tort law on balance tend to favor plaintiffs, *access* is difficult for plaintiffs. Taking into account both procedural and access rules, they argue that all things considered, plaintiffs do not have an unfair advantage. Black's proposal would increase the procedural barriers for plaintiffs tending, on the Gillette–Krier analysis, to shift the combined effect of access and procedural rules against plaintiffs.[135]

There is another point about victim access to the courts. It is difficult for the first group of plaintiffs with injuries from some toxic substance to recover, because the evidence may not be fully developed and there are no court precedents. The record of asbestos, DES, and other litigation suggests this. Once better evidence is available and legal precedents have been developed, tort recovery becomes easier.[136] Black's recommendations work particular hardships against "early" litigants, for there is unlikely to be the kind of peer-reviewed evidence he emphasizes. Yet there may be perfectly good evidence on which scientists can rely.

In addition, victims may not even bring cases in the first place, which weakens the deterrence aims of torts. Evidence from medical malpractice law indicates that "the fraction of medical negligence that leads to claims is probably under 2 percent."[137] This is important because medical patients are more likely to know who or what caused their injuries than are toxic injury victims. Thus, it appears that even in this area of tort law, where one might expect a high rate of suits brought, use of the tort law is not widespread and it achieves its "social objectives crudely."[138] We should expect more suits to be brought for malpractice than for toxic torts. Thus, we should not rush to restrict the tort law early plaintiffs even further by means of more stringent procedural rules in toxic tort cases.

The *Ferebee* rules were developed for circumstances "when the causation issue is novel and '*stand*[s] *at the frontier of current medical and epidemiological inquiry*.'"[139] When evidence of harm from toxic substances is at the frontiers of scientific knowledge, as is often the case, the tort law should not wait "until a 'statistically significant' number of people have been injured or until science has had the time and resources to complete sophisticated laboratory studies of the chemical."[140] The same Circuit Court of Appeals, however, also recognized the difference between a "frontiers of scientific knowledge" case and one in which there were 20 years of scientific studies to help decide the issue. In the latter case, *Richardson by Richardson v. Richardson-Merrell*,[141] the court upheld a trial court decision for the defendant because overwhelming scientific evidence favored it (although even in this case the issues may not be entirely clear-cut). Such flexible rules for coping with the evidence of possible harm from toxic substances are appropriate.

Furthermore, the position for which I have been arguing has also received some articulation from the Third Circuit Court of Appeals. In *In re Paoli R.R. Yard PCB Litigation*[142] that court overturned a district court's exclusion of scientific evidence in a case concerning alleged plaintiff exposure to polychlorinated biphenyls. The court viewed the Federal Rules of Evidence as embodying a strong and desirable preference for admitting any evidence having some potential for assisting the trier of fact and *for dealing with the risk of error through the adversary process*.[143]

A recent appellate decision by the Court of Appeals, Sixth Circuit, suggests a more specific rule for evaluating scientific evidence in a toxic tort suit.[144] That court upheld a district court verdict for plaintiffs who were exposed to hazardous substances (including carcinogens) leaking from a landfill. The district court decision was based on animal evidence and expert testimony. Moreover, the court declined to rule for defendants and rejected any reliance on an epidemiological study that had very low power (.25 to .30) to detect relative risks of 2 and 3, respectively. On this the whole court agreed. In a concurring opinion, however, Judge Jones suggested an approach to the sufficiency of scientific evidence that seems acceptable:

> It seems to me that the general standard of proof [to a reasonable medical certainty] can be met on a prima facie basis, where the record contains expert testimony on the chemicals and its properties (sic), and proof of exposure to the chemicals supported by some medical testimony. It seems to me at this point a rebuttable presumption of proximate causation ought to arise. The failure of the defendant to rebut this presumption would then give rise to a conclusion of reasonable medical certainty.[145]

Good animal studies, but no human epidemiological studies, and expert testimony supported the plaintiffs' case. There were no published human studies given any weight by the court, as Black often suggests the court must require.

Finally, the New Jersey Supreme Court recently developed rules for admissibility of scientific evidence in toxic tort suits similar to those articulated by Judge Jones of the Third Circuit Court of Appeals.

In *Rubanick v. Witco Chemical Corporation and Monsanto Co.*, the plaintiff's single expert witness, a retired cancer researcher at Sloan-Kettering Cancer Center in New York, testified that the plaintiff's decedent's cancer was caused by exposure to the defendant's polychlorinated biphenyls (PCBs). He based his opinion on (1) low incidence of cancer in males under 30 (plaintiff's decedent was 29), (2) decedent's good dietary and nonsmoking habits and the absence of familial genetic predisposition to cancer, (3) 5 of 105 other Witco workers who developed some kind of cancer during the same period, (4) " 'a large body of evidence' showing that PCBs cause cancer in laboratory animals," and (5) support in the scientific literature that PCBs cause cancer in human beings.[146]

> We hold that in toxic-tort litigation, a scientific theory of causation that has not yet reached general acceptance may be found to be sufficiently reliable if it is based on a sound, adequately founded scientific methodology involving data and information of the type reasonably relied on by experts in the scientific field. . . . [A causation theory that is not generally accepted must be offered by an expert who is] sufficiently qualified by education, knowledge, training and experience in the specific field of science. . . . The expert must possess a demonstrated professional capability to assess the scientific significance of the underlying data and information, to apply the scientific methodology and to explain the bases for the opinion reached.

This decision is quite appropriate given the recommendations made in this chapter. The court did not require epidemiological studies in support of the expert's opinion, but merely the appropriate "education, knowledge, training and experience in the specific field of science,"[147] and an appropriate factual basis for his opinion. The appellate division of the Superior Court (which apparently persuaded the Supreme Court) engaged in similar reasoning and appeared to be particularly sensitive to the problems of "early plaintiffs."[148]

These cutting edge cases all exemplify desirable trends in the tort law that permit tort plaintiffs to base their cases on expert testimony with appropriate factual support and not unnecessarily handicap their efforts, as Black's proposals and some court decisions would do. These trends should be encouraged to better serve the ends of tort law compensation and deterrence.

CONCLUSION

In this chapter I have argued against shifting the paradigm for scientific evidence offered in legal trials from the present procedure used in many jurisdictions— requiring proof to a reasonable degree of medical certainty—to a much more demanding one. This choice should be resisted, because it seems to impose a universal standard of evidence for quite different institutions, it risks distorting existing reasonable tort law relationships, and it errs on the side of exclusion rather than admission of evidence. Since the plaintiff bears the burden of production, requiring that this be met by proof equivalent to the criminal law's "beyond a reasonable doubt," as some have argued, substantially distorts the balance of interests between

plaintiff and defendant. And there are several defense strategies for countering
dubious evidence that do not require such a major change in the law.

This conclusion is a specific instantiation of the general principle that we properly should have different standards of evidence for different areas of human inquiry. There seems no good reason to subscribe to one (used in scientific fields) as
the proper one for all areas of human endeavor (for scientific, religious, legal, or
ordinary belief purposes). We certainly should not distort the tort law out of a
misguided belief that the scientific standards of evidence typically required in peer-
reviewed publications or demanded by the scientific community are the only appropriate ones for reasoning coherently about causal connections in toxic tort litigation.

3

Joint Causation, Torts, and Administrative Law in Environmental Health Protections

The Reagan Administration, in an intensive campaign to curb the role of the federal government, has eliminated or weakened scores of public health and safety rules, cut back enforcement programs and slashed services. . . .

Administration officials say their actions merely fulfill President Reagan's promise to bring efficiency and deregulation to government. A hands-off philosophy of regulation was mandated by the 1980 election campaign, in which Reagan crusaded to get the government "off the backs" of the American people, they argue.[1] *Los Angeles Times*, May 8, 1983

Smoking complicates the picture [of cotton dust causing brown lung]. The industry has long provided smoke breaks and smoking areas, but smoking seems to multiply the effects of cotton on the lungs. Many who get byssinosis also smoke, leading some doctors to say a disabled worker's damaged lungs are due only to smoking. Most Carolina lung specialists, however, say they see the disease frequently in nonsmokers as well.[2]

Robert Conn, Medical Editor
Charlotte (N.C.) *Observer*

The entire proposition that "byssinosis" is a crippling disease caused by cotton dust is unfounded and impossible, scientifically, to prove. Burlington industries . . . ran an exhaustive study on this subject that took years and cost millions of dollars. The conclusive results of testing of thousands of employees in cotton mills was that less than 5 percent had respiratory problems, and over half of these were heavy smokers. The doctors concluded that whatever ailed this small percentage was neither a crippling disease nor an illness but a discomfort.

This is simply not a fight against so-called "byssinosis" or "brown lung." This fight is as old as civilization: The unending war of a free people with inalienable rights granted by God, against those tyrannical power-hungry politicians intent on the establishment of a totalitarian government.[3]

W. B. Pitts, President
Hermitage, Inc.

The inability to distinguish lung impairments which are caused by smoking, cotton dust or a combination of the two has wreaked havoc on the workers' compensation process. Individuals with obstructive lung disease and a history of exposure to cotton dust are now often presumed to be suffering byssinosis even if they have smoked two packs of cigarettes a day for 40 years. Because symptoms of classical byssinosis are no longer considered essential to the diagnosis, we are left picking up the tab for lung impairments caused by factors other than cotton dust, primarily smoking. That can be a huge tab when you remember that 20 percent of the population suffers from lung ailments. The industry should not be forced to pay for something that is not its fault.[4]

Dr. Harold R. Imbus, Director of Health and Safety
Burlington Industries, Inc.

The preceding comments summarize part of a recent controversy concerning the need for regulating hazardous substances in the workplace. Regulatory law in general as well as environmental and workplace health law has been increasingly under attack. Some criticisms are a matter of political philosophy and objections in principle to governmental intervention: a presidential administration that held that the less government the better, a traditional American attitude opposed to any coercive interference with the marketplace (as W. B. Pitts stated). I do not take up these political philosophical issues, although I consider some reasons that bear on them.

Objections to regulation, which may be widely held, are also suggested; these rest upon claims about causation or upon views about the proper principles for allocating liability for causation. A species of such objections—the problems of joint causation—is the focus of this chapter. If these objections were correct, they would undercut some of the efficacy of both tort and administrative health law.

There are circumstances in which a person is exposed to two or more substances each of which might be sufficient by itself to cause the disease in question or circumstances in which one substance in conjunction with other conditions might produce the disease. Because toxic substances cause harm "by the unfelt intrusion of invisible molecules,"[5] unless those molecules leave a unique signature, which is rare,[6] there is great difficulty in determining the cause of injury, as we have seen. Furthermore, one may well have been exposed to other invisible intruders that might have caused or contributed to the same injury. Even worse, one's own behavior (smoking) or genetic makeup may have been a factor.[7] These problems could plague both torts and administrative law.

There are, of course, a number of differences between the two areas of law. The tort law sets public standards of conduct that are privately enforced, typically by the injured party, with ex post remedies for injuries suffered.[8] By contrast, the regulatory law, which issues protective regulations prior to injury in order to prevent harm,[9] is typically publicly enforced by a government entity. (We consider other differences later.)

For both areas of law the causal claims needed to prove harm from toxic substances are often difficult to establish. We have already seen some of the scientific reasons for this; when joint causation obtains the difficulties of proof are even greater.

A more sophisticated objection is that even if such causal claims can be established when joint causation obtains, it is not permissible to impose tort liability or preventive regulation on firms because either contributory cause might be the culprit. Because protective health laws have been under attack, I examine possible objections to imposing tort liability or administrative regulation on firms when joint causation obtains.[10]

This chapter explores traditional and well-established principles of liability for causation taken from the tort law. They reflect deeply held, widely applied, and long used principles for compensating persons for injuries inflicted on them.[11] I draw on cases from tort and worker's compensation law not to replicate either exactly, but to extract and examine some common principles of liability for causation taken from both contexts.[12] This strategy, similar to one articulated by Ronald Dworkin, aims to discover philosophical principles underlying decisions in particular areas of the law as well as those that permeate the law more widely.[13]

The principles I discuss show that there is no obvious legal bar to liability when joint causation obtains in tort and workers' compensation law. Furthermore, courts appear to have endorsed the joint cause rules in toxic tort cases, and these hold some promise for extending the law. These results suggest in turn that there should be no legal bar to administrative regulation simply because joint causation obtains. Moreover, many of the same principles that justify imposing liability in torts would also justify preventive regulation of toxic substances in the workplace. The result of this inquiry is that administrative agencies have a comparative advantage over the tort law to prevent harm from toxic substances when joint causation obtains principally because causation is easier to establish for regulatory than for tort law purposes.

In principle these and other considerations are good reasons for using administrative agencies to provide protection from carcinogens. Despite the theoretical advantage, however, agencies and their regulations should supplement, not replace, the tort law, because of the political problems that plague them.

JOINT CAUSATION AND THE TORT LAW

Liability Rules for Causation

Assuming that causation can be established when there are joint causes, would there be any legal bar to imposing tort liability on a firm that has exposed an employee to a hazardous substance simply because joint causes exist?

The thesis that when two or more causes contribute to a harmful result the agent of neither cause should be held accountable seems to receive its sanction from one of the most fundamental rules of the tort law. This is the ''but-for'' rule:

> The defendant's conduct is a cause of the event if the event would not have occurred but for that conduct; conversely, the defendant's conduct is not a cause of the event, if the event would have occurred without it. As a rule regarding legal responsibility, at most this must be a rule of exclusion: if the event would not have occurred ''but for'' the defendant's negligence, it still does not follow that there is liability, since other considerations remain to be discussed and may prevent liability.[14]

It may be some rule like this that Dr. Imbus has in mind in trying to exclude cotton mills from workers' compensation liability. He argues in effect that many who suffer from chronic bronchial obstruction have contributed to their own harm by smoking. The emphysema, he suggests, would have occurred even if the employee had not been exposed to cotton dust. His intuition and the but-for rule make sense, for if something else would have caused injury to an employee, even in the absence of exposure to cotton dust, it seems the mill should not be held accountable for such injury and should not be regulated to prevent such injuries in the future.

Despite this seeming plausibility, the tort law permits recovery in many similar cases. Again Prosser and Keeton say:

> If two causes concur to bring about an event, and either one of them, operating alone, would have been sufficient to cause the identical result, some other test is needed.[15]

That "other test," first provided by a Minnesota court, has found general acceptance:

> The defendant's conduct is a cause of the event if it was a material and a substantial factor in bringing it about. Whether it was such a substantial factor is for the jury to determine, unless the issue is so clear that a reasonable person could not differ.[16]

> If the defendant's conduct was a substantial factor in causing the plaintiff's injury, it follows that he will not be absolved from liability merely because other causes have contributed to the result, since such causes, innumerable, are always present. In particular, however, a defendant is not necessarily relieved of liability because the negligence of another person is also a contributing cause, and that person, too, is to be held liable. Thus where two vehicles collide and injure a bystander, or passenger in one of the cars, each driver may be liable for the harm inflicted. The law of joint tortfeasors rests very largely upon recognition of the fact that each of two or more causes may be charged with a single result.[17]

The substantial cause rule has been instantiated in several variations which present distinguishable subcases especially relevant to environmental and workplace health hazards caused by toxic substances.

The first case (where the first formulation of the rule appeared) concerns the liability of the Minneapolis, St. Paul and Sault St. Marie Railroad Company. A spark from one of the railroad's engines caused a forest fire that combined with another fire of unknown origin and swept over Anderson's property. The Supreme Court of Minnesota upheld the following jury instructions at the trial court level:

> 1. If you find that other fire or fires not set by one of the defendant's engines mingled with one that was set by one of the defendant's engines, there may be difficulty in determining whether you should find that the fire set by the engine was a material or substantial element in causing plaintiff's damage. If it was, the defendant is liable, otherwise it is not.[18]

Call this the "negligent–innocent" cause rule, for both a negligent and an innocent cause may have combined to produce the damage to Anderson. This has obvious application to environmental and workplace health hazards, for even if a person is exposed to a naturally occurring substance or suffers a genetic change which causes a disease such as lung cancer, if he was also negligently exposed to a toxic substance in the workplace which causes the same disease, he can recover, provided causation (and the other elements of a tort) can be established.

Still other cases permit recovery where an actor's negligent conduct precipitates a preexisting condition. In *Steinhauser v. Hertz Corp.*,[19] the Steinhauser car collided with a rental car owned by Hertz Corporation. The Steinhauser daughter apparently suffered no bodily injuries in the accident, but she began to suffer episodes of schizophrenia shortly after the accident. She was finally diagnosed as a schizophrenic and it was predicted she would be under the care of a psychiatrist and intermittently institutionalized most of the rest of her life. The California Supreme Court held:

> 2. When a defendant's negligent conduct *triggers* or *precipitates* a preexisting condition which might have developed anyway, defendant can be held liable, although a jury hearing such a case may want to take the preexisting condition into account in awarding damages.[20]

Call this the "precipitation" rule, for it illustrates the principle that when a defendant's faulty conduct precipitates a preexisting natural condition, the defendant can be held liable. This rule also has obvious application to workplace and environmental exposures to toxic substances. If a person is genetically prone to emphysema or cancer and negligently exposed to an environmental toxin, then on this principle he would not be barred from recovery for the toxic exposure, provided the appropriate causation and other elements of a tort could be proved.

In *Evans v. S. J. Groves and Sons Co.*,[21] the U.S. Court of Appeals, Second Circuit, discussed several cases related to but somewhat different from *Steinhauser*. That court identified the "acceleration" rule:

> 3. When a plaintiff, "already incapacitated in some degree by a disease or injury, suffers a worsened condition as a result of the defendant's wrongful act the defendant is liable for the additional harm that he caused."[22]

Thus, when a defendant faultily "accelerates" an existing disease or injury, he may be held accountable for the resulting additional injuries. This rule, too, has obvious applications to environmental exposures to toxic substances. If a person already suffering emphysema has it accelerated by exposure to a substance such as cotton dust, there would be no bar to recovery, provided causation (and the other elements of a tort) could be proved.

The next class of cases exhibits what might be called the "special condition" rule:

> 4. When a plaintiff with a special physical condition perhaps in no way permanently disabling, such as pregnancy, which makes the consequences of the negligently inflicted impact much more serious than they would be for the normal victim, the defendant is liable for the entire damage unless he succeeds in establishing that the plaintiff's preexisting condition was bound to worsen, in which event an appropriate discount should be made for the damages that would have been suffered even in the absence of the defendant's negligence.[23]

This has obvious application for exposure to toxic substances, for if a person has bad lungs from birth or from childhood disease and is negligently exposed to a substance which worsens his condition, then there is no bar to recovery of compensation for the worsened condition, provided causation (and the other elements of a tort) can be established.

Next, there are cases close to those that concern Dr. Imbus. *McAllister v. Workmen's Compensation Appeals Board* concerns a fireman of 32 years' standing who died of lung cancer. His heirs sued the city of San Francisco for injuries arising out of and in the course of his employment. The interesting facts about this case are (1) that the decedent had smoked about a pack of cigarettes a day for some 42 years, although he did not inhale during at least part of that time, and (2) in the course of his employment he had been repeatedly exposed to smoke that contained the same kinds of toxic substances and carcinogens as cigarette smoke itself.[24] The court permitted recovery in this case and seemed to subscribe to the following principle:

> 5. When a plaintiff's employment contributes substantially to the likelihood of his acquiring lung cancer, even though his own conduct increases the risk of the same disease, e.g., because of smoking, as long as his risk of contracting the disease by virtue of his employment is materially greater than that of the general public, then he may recover worker's compensation for injuries suffered in the course of and arising out of his employment.[25]

Call this the "additive cause" rule, for both the employee's workplace exposure and his own personal habits contribute to his acquiring a serious disorder. The application of this principle is obvious.

Finally, consider what we may call the "weakening cause" rule. In *Selewski v. Williams*[26] the plaintiff and his son were injured in an automobile accident caused by the defendant. After being treated in the hospital for various minor physical injuries and remaining overnight, the plaintiff was released, but his son remained in the hospital. Plaintiff spent many hours watching over and worrying about his son's condition. About two weeks later he came down with pneumonia. The trial court jury found that the plaintiff's illness was caused by the defendant's collision with the plaintiff's automobile. Upon appeal the Court of Appeals of Michigan held:

> 6. The jury could have found the appellant suffered a slightly lowered vitality resulting from severe aches and pains which prohibited him from achieving normal and restful sleep. This factor, coupled with the appellant's anxiety, irregular eating habits, and hospital vigil due to his son's injuries was sufficient for the jury to have found that the appellant's weakened vitality either diminished his capacity to withstand the virus, or caused the virus to become more virulent.[27]

Thus, here is a case in which a negligent actor is liable for injuries which follow proximately from the initial injury. There are many variants on this in the tort law, but one is sufficient for our purposes. The analogue for exposure to toxic substances is clear: if exposure to a toxic substance makes a person less able to resist other diseases, then a defendant faultily exposing him to such substances may be held liable for further diseases (provided all the other elements of a tort can be established).

Given the preceding rules there is no difficulty in principle, even in the tort law, recovering damages for harm suffered when two or more causes contribute to a result or when a human cause and an existing background condition combine to produce a harmful result.

These general principles not only apply to traditional tort cases but also have been endorsed in some jurisdictions for use in toxic tort litigation.[28] The court in *Elam v. Alcolac* held:

> Thus, although our law requires proof of cause to recover in tort, it does not require proof of a single cause. The substantial factor standard . . . is particularly suited to injury from chronic exposure to toxic chemicals where the sequent manifestation of biological disease may be the result of a confluence of causes.[29]

And the *Alcolac* court required the plaintiff to show "that the Alcolac conduct had such an effect in producing the harm as to lead reasonable men to regard it as a cause."[30] This rule, endorsed by the Restatement (Second) of Torts,[31] is ambiguous. It fails to distinguish whether the law requires that defendant's conduct (1) causally contributed to the plaintiff's *harm* or (2) causally contributed to increasing the plaintiff's *risk* of harm. Instead the court leaves the rule as an *epistemological* one: (3) whether reasonable people considering the effect of the defendant's action *could regard* it as a cause.

In reviewing the evidence and the lower court decision, however, the court's view seems clearer. First, it notices that causation rules "interdepend" with other elements of a tort offense to deter wrongful conduct, to see to it that innocent victims of others' conduct will receive redress, and to see to it that the costs of exposure to toxic substances "shall be borne by those who can control the danger and make equitable distribution of the losses, rather than by those who are powerless to protect themselves."[32] Second, the burden of proof must be "accommodated to the quality of evidence the scientific community deems sufficient for that causal link."[33] Third, the court permitted plaintiffs to rely upon circumstantial evidence as to the substances emitted.[34] It also permitted reliance upon animal studies that the substances could have caused the harm in question, for at high doses they are known to cause such harms in animals, and this is about the only evidence it had available.[35]

These considerations were sufficient for the appellate court to refuse to overturn a jury verdict. This suggests that the appellate court probably considered the evidence sufficient if it increased the *risk* of the kind of harm the plaintiff in fact suffered.

This interpretation may extend the substantial cause rule, but even this is not clear, for it is not substantially different from the *McAllister* rule noted previously. Nonetheless, it appears quite appropriate for toxic torts, given the kind of evidence that is usually available and the aims of the tort law (recited by this Missouri appellate court). Furthermore, while this will make it somewhat easier for plaintiffs to recover, it is not clear that it will upset the balance of interests in tort suits.[36]

From the preceding discussion, it appears there is no legal bar to establishing a defendant's tort or workers compensation liability for faultily making a substantial and material contribution to a plaintiff's injuries, provided all the other elements of a tort law action can be shown. The most recent toxic tort cases (*Elam v. Alcolac*) suggest that a plaintiff may be permitted to establish a case on the grounds that exposure to a toxic substance increased the *risk* of harm which in fact was suffered.

Perhaps, however, some of the unease about joint causation comes from concern that the causal claims cannot be established. Again, we find these beliefs mirrored or perhaps writ large in the causation rules of the tort law.

Proof of Causation

In general one must establish each element of a tortious act with proof by a preponderance of the evidence.[37] We have seen that this would require something over 50% of the evidence to establish each element of a tort.[38] However, in proving cause-in-fact, the standard may not be quite so demanding. As Prosser and Keeton put it:

> The plaintiff is not, however, required to prove the case beyond a reasonable doubt; it is enough to introduce evidence from which a reasonable person may conclude that it is more probable that the event was caused by the defendant than that it was not.[39]

If we use the language of "substantial cause" to cover all cases of joint causation, we might say that the plaintiff must show that it is more probable that a defendant's conduct was a substantial cause of a plaintiff's injuries than that it was not. This rule, as we have just seen, may be extended as it was in *Elam v. Alcolac* to permit recovery if the court finds that the defendant's actions increased the risk of the kind of harm which the plaintiff suffered.

A major problem arises in applying this proof rule when two conditions obtain: (1) the disease produced by each of the causes cannot be identified by the nature of the cause, that is, a particular substance does not leave a unique "signature" on the disease (we might call this a nonunique cause), and (2) the evidence for the cause is statistical in nature, as is the case with most diseases caused by exposure to toxic substances.

We can illustrate these points with respect to diseases caused by exposure to radiation. Exposure to radiation in power plants or in nuclear processing plants may cause leukemia in those so exposed. However, leukemia is also a naturally occurring disease afflicting about 10 out of every 100,000 people.[40] Suppose workers in a plant are exposed to radiation sufficient to double the leukemia rate among those in the plant compared to the general population—the so-called doubling dose. The disease rate among those exposed would then be about 20 cases of leukemia per 100,000 people. If a particular person who has worked in a nuclear power plant has leukemia, one could not tell from the disease whether it was caused by exposure to radiation or was a naturally occurring disease that would have been acquired in any case. The radiation leaves no "signature" on the disease so that one can identify the particular cause. (We should note that there are additional problems if a particular worker had been exposed both to radiation and to concentrations of benzene great enough to cause leukemia, for then we have triply joint causation: two possible workplace causes and a possible naturally occurring cause.)

In addition, if the evidence for causation is statistical, then proof is even more difficult. For example, suppose that Sam, who worked in a nuclear power plant for 20 years, became ill with leukemia as a result of the radiation exposure, and suppose that he was exposed to something less than a doubling dose of radiation during his employment (suppose that exposure was sufficient to cause a leukemia rate of 16/100,000). On these facts alone he probably could not recover compensation for injuries suffered according to standard tort law rules because he could not prove that his injuries more likely than not were caused by exposure to radiation. If his exposure were sufficient to change the disease rate from 10/100,000 to 16/100,000, only 6/16 of the cases, or 37%, would be attributable to the radiation.[41] Such statistics are insufficient to meet the tort law standard of proof. Thus, no one, including Sam, would recover compensation, including those whose leukemia was caused by radiation exposure. On the other side of the same issue, if employees were exposed to more than doubling doses of radiation, all who contracted leukemia, including the half of the cohort whose leukemia was naturally caused, would be entitled to recover compensation. This result would be unfair to defendants. Given such proof requirements, and given that causal claims in most cases will have to be established by means of statistical evidence, it will be difficult indeed to establish a claim in tort law for particular plaintiffs for compensation for injuries suffered.[42]

As a consequence of these as well as several other problems, commentators have argued for use of regulatory law to govern legal relations in this area rather than, or in addition to, private tort law.[43] This seems to me correct in principle and I argue for it later. Before considering that issue it is important to understand some of the main differences between the tort and administrative law and some of the theoretical advantages of the latter. This is especially true for proof of causation.

ADMINISTRATIVE LAW

As we saw in the introduction to this book there are several differences between tort law and administrative law. The tort law primarily provides privately enforced ex post remedies, whereas regulatory law provides an ex ante publicly enforced remedy. In applying the tort law, then, judges frequently rely upon backward-looking reasons of compensation (and in some cases retribution) to justify imposition of liability, whereas regulators rely upon forward-looking reasons of prevention to justify imposition of regulations on a party. Despite the differences, the issues are not quite so sharply drawn. Often in tort cases, courts in making appellate decisions appeal to reasons of deterrence or prevention of harmful conduct and to reasons of distributive justice as well as to reasons of compensation and corrective justice. These general principles can guide administrative policies as well.

No Moral or Policy Bars to Regulation
When Joint Causation Obtains

The common law of torts has evolved in such a way that there are no legal grounds for denying compensation to a victim simply because joint causation obtains. Indeed tort law cause-in-fact rules have evolved in order to accommodate the more complicated facts of joint causation. Similarly, I think that there are no obvious moral or policy bars to using regulatory law when joint causation obtains.

To see this we need some additional terminology to clarify the notion of causation. Joel Feinberg has noted that there are various "contexts of inquiry in which causation citations are made": explanatory contexts, in which we seek to explain causally why something happened, engineering contexts in which the aim is to facilitate control over future events by citing the most efficiently and economically manipulable causal factor, and blaming contexts in which the aim is to fix blame or responsibility for causation.[44] We might think of the tort law as largely concerned with attributing blame to human agents in order to compensate victims for harms suffered. The tort law is also concerned with singling out agents who can best prevent the harm or best afford the compensation owed a victim, more of an "engineering" strategy.[45] Administrative law on the other hand might be seen primarily as taking an engineering perspective within the scope of legal remedies to prevent harm. However, even if we take this preventive viewpoint, not only is there no obvious bar to imposing regulations when joint causation obtains, but administrative agencies have a comparative advantage over torts for preventing environmental health harms.

First, if we consider the rationales behind the "precipitation," "acceleration," "special condition," and "additive cause" rules, none would bar use of administrative remedies when joint causation exists. The principle underlying all of them seems to be that if a defendant faultily makes a material or substantial causal contribution to a victim's injuries, even if his contribution is not the major one, he is nevertheless held accountable to pay some compensation for this faulty contribution. (However, for most of these rules there is the further and separate question of whether a negligent defendant should be required to pay for all of a victim's injuries, or only for those that he causally contributed to.) This principle seems to apply directly to the preventive aims of regulatory law: if a firm is in fact making some causal contribution to the deaths and diseases of its workers, and even if the firm's behavior is not faulty, *ceteris paribus,* there is still a case for regulations to try to control such harms. If one can reduce or eliminate a substantial or material causal contribution to employees' ill health by means of appropriate regulations and engineering, it is surely important to do so.[46] How extensive and costly such regulations should be may be a matter of some debate, just as the amount of damages is a matter of some debate in the tort law, and such matters would have to be decided on the facts of the particular case, the costs, and so on. But the mere fact of multiple possible causes of harm should not bar consideration of administrative rules to protect potential victims.

Next, if it were discovered that all employees exposed to a particular toxic substance, for example, cotton dust, were thereby so weakened that they could not resist diseases or that they contracted various lung disorders, it seems that on analogy with the weakening cause rule preventive regulation would be defensible. For if such illnesses could be reduced or eliminated by preventive measures, there is good reason based on these principles for doing so. Again, the nature, extent, and cost of regulation would have to be addressed separately.

What is common to all the cases of joint causation we have considered under the precipitation, acceleration, special condition, additive cause, and weakening cause rules is that an agent's (or firm's) behavior contributed, along with other causes and conditions, to a victim's harm (or risk of harm). Surely if epidemiological studies reveal such patterns, then, *ceteris paribus,* there is a case for regulations. A firm's contribution "makes things worse" for the victim than they otherwise would have been. If preventive regulations will make things better for employees, there is surely no policy bar to regulating in such circumstances. (Whether they should actually be required if the firm's contribution is a minor one is another issue.)

Finally, the only rule we have yet to discuss is the negligent–innocent cause rule. The principle underlying the negligent–innocent cause rule seems to combine epistemological concerns with a presumption of the tort law. One consideration seems to be the epistemological one that a court might not know or have any evidence as to which of two causes, both sufficient (together with other conditions) to cause the harm in question, in fact caused the damage. A second consideration is that the victim is certainly innocent and the defendant certainly acted faultily, even though his faulty behavior may not have *caused* the harm in question. Hence, we have the victim's innocence, combined with the defendant's behavioral fault, but possible "causal innocence." Thus, the principle seems to be that it is better that a certainly negligent actor who possibly did not cause the harm (but who might have caused the harm) be made to pay compensation than that a certainly innocent victim be made to suffer injuries without compensation. The principle rests partly on a fault rationale, partly on an aim of the tort law, and partly on epistemological grounds. One might see this as an extreme case of a decision under uncertainty: one alternative risks mistakenly holding a faulty but possibly "causally innocent" defendant accountable in order to compensate a certainly innocent victim; the other alternative risks not compensating a certainly innocent victim and exonerating a certainly faulty but possibly "causally innocent" defendant. The analysis in Chapters 2 and 5 suggests this might be the appropriate decision. Although this principle implies that in the tort law avoiding a false negative is of greater importance than avoiding a false positive (at least for a negligent act) in absence of all other information, I did not rely on this assumption in Chapter 2.

This particular principle does not transfer especially well to preventive administrative law until some factual evidence is provided. If a regulatory agency does not know or have evidence of what might cause a disease like leukemia, a natural cause or exposure to a toxic substance like benzene, it cannot rely upon the fault of one party (as can be done in the tort law). It can call on a presumption of most environmental health laws (that the public health be protected).[47] However, when a regulatory agency has some general statistical evidence that a substance increases morbidity or mortality (that is identical to diseases produced naturally), this evidence removes the epistemological uncertainty and tends to confirm a firm's contribution to harm, even if not the firm's fault. Further, if the appropriate goals of regulation were supplied, for example, by the Occupational Safety and Health Act, such goals provide normative analogues to the presumptions of the tort law. An administrative agency would not rely on a firm's "fault" but on the preventive goals of its enabling legislation. Once these two claims are established, the argument for regulation would be plausible.

This completes the argument that just as there is no legal bar in the tort law to imposing compensatory liability on an agent merely because joint causation obtains, there is no obvious policy bar to imposing preventive regulations on a firm in the same circumstances. The remainder of this chapter discusses some steps toward establishing a stronger conclusion, namely, that there are good reasons drawn from tort law and applied to the regulatory context for imposing preventive regulations on firms.

The Case for Regulation of Toxic Substances When Joint Causation Obtains

Despite obvious differences between tort and administrative law discussed earlier, one might think that if conduct is properly compensable in torts, it is properly preventable by regulatory law. That is, with respect to the specific issue of joint causation one might think that if there is no legal bar to tort law compensation when joint causation obtains, there should be no bar to using regulatory law to try to prevent identical harms. Although I think there is much to this idea, the inference is not automatic.

First, there are numerous acts for which tort law remedies are readily available, but which are not thereby made the subject of regulatory law; these acts range from automobile accidents to my tree falling on your house to my spewing cement dust on your land. There are many areas of our social life governed by tort law that are not appropriately governed by administrative law, for torts as an institution have advantages over administrative law. This is the case for the examples just considered.

The tort law tends to have a comparative advantage over administrative law in a number of circumstances. Two of the most important are where an ex post "remedy" or rule—one "triggered only by the occurrence of harm"—is appropriate and where *privately* initiated (vs. state-initiated) remedies are appropriate.[48] Retrospective remedies are superior to ex ante approaches (which apply before, or at least independently of, the occurrence of harm)[49] where the administrative costs will be much higher with an ex ante approach. Private remedies are superior when a victim's information about a risk or occurrence of harm is likely to be superior to that of a public agency's.[50] Victims are likely to possess superior information to a public agency when they are clearly aware of having suffered the harm, know the identity of the responsible party, and have observed the responsible party's behavior in circumstances typical of ordinary tort cases.[51] Finally, where "injurers' assets are not small in relation to the harm they may do and injurers will not often escape responsibility for harm" the private, ex post remedy of the tort law will have an advantage.[52] When these obtain there is quite a weak case for administrative regulations to accomplish the same ends. Conversely, there are circumstances in which administrative remedies possess substantial advantages over torts. I return to some of these issues later.

Next, some reasons that justify use of the tort law in particular cases cannot justify use of the administrative law. In particular, justifications that appeal to an agent's faulty behavior are relied upon in torts but seem inappropriate in regulatory law because the aim of regulation is to prevent harms before they occur, possibly before an agent has had the time to exhibit faulty conduct, possibly even before an agent is *aware* or should have been aware of the harmfulness of his activities. Appeals to an agent's faulty conduct may be thought to justify principles which prevent unjustifiable gains and losses (the latter principles rest upon the more general moral principle that those at fault should forfeit any gains made from faulty conduct or the principle that one ought not to profit from one's own wrongdoing).[53] Finally, use of regulatory law may be thought to be more costly. Administrative agencies need an organizational structure to investigate the need for regulation, to formulate regulations, and to enforce compliance. And with regulation would come the costs of compliance for the regulated industries.

Despite these reasons for thinking that justifications for using the tort law do not automatically transfer to justifications for using regulatory law, there are reasons of both utility and justice drawn from the tort law for using regulatory law in addition to the tort law to regulate use of hazardous substances when joint causation obtains.

Utilitarian Arguments

First, it is much easier to establish causal claims needed for regulation than to establish causal claims typically needed for a tort suit.[54] We have seen that one shortcoming of torts is the difficulty of establishing that a particular defendant's conduct *caused* a particular plaintiff's injuries. But for ex ante regulation one need only show *a general statistical connection* between exposure to a hazardous substance and contraction of a disease.[55] Consider the radiation cases discussed earlier. Once it is known that a certain dosage of radiation results in doubling of the rate of leukemia among those so exposed compared with the general population, providing there are no confounding factors involved (and the epidemiological study makes biological and medical sense), this is all that may be required for regulations such as those imposed by OSHA.[56] However, such causal evidence is insufficient to establish the particularized causal connections needed in torts. In addition, as we saw in discussing the additive cause rule from *McAllister,* the court recognized that there could be statistical evidence showing that it was dangerous to one's health to work in a particular occupation such as firefighter, yet there might be insufficient evidence to show that one's particular injuries were caused by working in that occupation. This is not to say that it will be easy to provide such statistical or epidemiological evidence. Indeed as we saw in Chapter 1, this may be especially difficult to establish.

In addition to ease-of-proof considerations, there are forward-looking utilitarian arguments which justify compensation in the tort law and which similarly can justify using regulatory law to prevent harms caused by exposure to toxic substances when joint causation obtains. The first is what we might call a "costs to individuals" argument. If we ask who is better able to bear the costs of regulation—employees (and their families) who would contract diseases and on whom the costs would fall were there no regulations or industries on whom the costs would fall were regulations required in order to try to prevent such diseases—the answer seems clear. In the absence of regulations, the costs of disease from exposure to hazardous substances would fall heavily on the victims and their families. The costs to individuals in this group of people would likely be enormous. If there were regulation of a firm that utilized toxic substances, there would indeed be the costs of prevention, but the firm can spread such costs intertemporally and interpersonally through the prices of its products, through somewhat lower dividends to its shareholders, and through insurers.[57] The costs per affected individuals in that case would tend to be quite small.[58] Unless there are costs of comparable magnitude to individuals who would be deprived of products as a result, this argument greatly favors regulation. (Although the number of individuals affected is pertinent.)

A second utilitarian argument is a deterrence or cost avoidance argument. The costs of regulating risks or not should be imposed upon the party to the controversy that is better able to avoid the costs that might result.[59] The firm that would introduce toxic substances is in the best position to avoid costs that might be caused by introducing them, for presumably it has greater knowledge of the risks involved and is better able than employees to avoid the risks by not imposing them. Firms introducing such substances presumably know the nature of the substances introduced, whether they are structurally similar to known toxic substances, whether they are toxic to mammals, and so on. Workers, on the contrary, know next to nothing about such substances.[60] The general public is typically even more poorly informed about such matters. Thus, based upon this argument, it is much better to place the costs of regulations on firms that would introduce hazardous substances than to place the costs of disease or death on employees, the public and their families.

The third argument is related to the second. The previous argument assumed that a firm might know that a particular substance posed some risk, even a small one, of causing disease or death in employees exposed to it, but frequently this is not true. When neither firms nor employees are aware of risks from a substance, who should bear the costs of such uncertainty? There is another way of phrasing this question: Who should bear the risk of ignorance? Who is better able to resolve uncertainty about toxic substances—firms that introduce potentially toxic substances or workers who may be exposed to them? The answer appears to be the former.[61] In the regulatory context, under a postmarket regulatory statute workers and the public typically bear the costs of ignorance, but are not in a position to remove it.[62] Under a premarket regulatory statute, a firm would not be permitted to introduce a substance into commerce if there were considerable uncertainties as to its toxicity. This is the way several premarket statutes presently work, including sections of the Food, Drug and Cosmetic Act as well as the Federal Insecticide, Fungicide, and Rodenticide Act.[63]

The last two arguments emphasize a "knowledge-generating"[64] effect that the imposition of tort liability or regulatory standards will likely have on the behavior of firms. Given the many unknowns surrounding a firm's products and processes—that is, a substance's toxicity, mode of transmission, and possible effects on persons (if any)—imposition of such standards on firms may encourage firms (1) to do research into injury causation, (2) to investigate the manner in which their actions endanger others, and (3) in general to make the relevant technologies "safer and more humane."[65] Furthermore, an administrative agency, by imposing or being willing to impose regulations on firms when it is demonstrated that excess harms arise from employee exposure to hazardous substances, thereby places incentives on the firms to acquire information needed to show that a workplace is not hazardous, and to do the research required for new technologies to provide safer processes and products. All of this information generation seems salutary from a utilitarian point of view and would not be achieved by leaving employees unprotected and letting the losses lie where they fall.

We have thus far considered, in addition to an ease-of-proof-of-causation argument, three major utilitarian arguments traditionally used in the tort law which provide equally good or better reasons for imposing preventive regulations on firms whose products or processes may be hazardous to its employees. An important point to notice about these arguments is that although they come directly from the tort law, they apply quite appropriately to regulatory law. Analogous arguments would establish an administrative rationale for general environmental health protections. The causation argument provides a decisive advantage for regulatory over tort law.

Justice Arguments

As indicated, in shifting the discussion from torts to regulatory law, we lose access to what is thought to be one of the major justice arguments available to tort law justifications: a defendant, by faultily and wrongfully invading a victim's legal "space" in which he is entitled to remain free from interference and harm, is, as a matter of corrective justice, properly compelled to repair the damage he has done. Of course, this rationale is not available to justify forward-looking regulations, since one cannot assume or presume that all firms that should be regulated will have wrongfully invaded their employees' rights. Nevertheless, this caveat aside, there are some arguments of fairness and distributive justice pertinent to justifying preventive regulation.

A major consideration of justice, or perhaps fairness, is that the party who realizes benefits from imposing especially hazardous risks on others should take precautions to see to it that the risks do not materialize. This is the analogue of the tort rule that the costs of an activity should be borne proportionally by those who benefit and benefit most from it.[66] If in the regulatory context this principle would require that those who stand to benefit most from an activity should be required to take precautions themselves, then there is a reason for imposing regulations on them which require them to take the appropriate preventive precautions according to this principle.

The rationale is that in an enterprise those who benefit most from it should be identical to those who bear the sacrifices to make it possible and that the sacrifices for each should be proportional to the benefits each receives.[67] Of course, actual harms to victims from an activity producing or using hazardous substances are the costs of the activity. Risks of harm, however, are also costs of the activity.[68] Even if an employee works for a firm and benefits from his employment, this does not remove the onus of the fairness argument—that the firm should bear the cost of preventing harms from toxic substances befalling an employee. His benefit appears minuscule compared to those who own and operate the firm and his costs, should he die or suffer serious illness, would be quite large indeed.

A second principle is that if one intentionally exposes another to great danger, however socially desirable the activity, one should bear the risk of loss to the person, should it occur, and the costs of taking precautions to see to it that such losses do not occur. This principle is analogous to one in the tort law used to justify imposition of strict liability on those engaged in dangerous activities. The rationale behind this principle is not entirely clear. One consideration suggests that it would fall under one of the utilitarian rationales already considered.[69] A second consideration seems to be that the world should be free from ultradangerous activities; agents who engage in them, even though they are on balance beneficial, should bear any costs if harm materializes. Thus, in the tort law such agents are held accountable without proof of fault to redress any adverse consequences of their activities which have unavoidable ultrahazardous risks.[70] This consideration would seem to justify preventive regulations as well.

If the introduction of ultrahazardous activities into the community is sufficient to justify imposing strict liability on the agent, then surely the introduction of ultrahazardous activities should justify, *ceteris paribus*, the imposition of preventive regulations to minimize the possibility that such activities harm others. If ultrahazardous activities are of sufficient concern to merit strict liability in torts, this suggests they may be permissibly regulated.[71]

To this point I have argued that several general tort law rationales which would justify tort liability easily transfer to regulatory law and justify the imposition of preventive regulations in the case of joint causes that pose serious health hazards. Furthermore, we have seen how the general principles underlying the specific joint cause rules of the tort law also justify imposition of regulatory law. (At least they do not bar imposition of regulations, and the same arguments would also provide reasons for regulations.) As a penultimate point in the argument, it is important to notice that a person's health is one of his most important goods, for according to the leading ethical theories and workplace health protections it would seemingly have high priority. This point is developed further in Chapter 5.

If we combine the foregoing considerations with the fact that causation is much easier to establish for preventive regulatory purposes than for tort law purposes, we have very good reasons for using administrative remedies when joint causes obtain. Since protecting people's health is one of the most important aims of the leading moral theories (and protecting workplace health seems equally important), and since one would want to reserve liberty-restricting regulations for securing the most important moral or policy aims, these reasons strengthen the general case made here for using regulatory law to protect employees' health in the workplace.

Before closing this section, we should note a final point about costs. Normally, as we discussed earlier, it is thought that use of administrative measures is especially costly (because of the need for a permanent bureaucracy, etc.) and certainly more costly than relying upon the tort law to accomplish the same aims. Without considering other cost issues, however, it may be that the costs associated with proving causation can be quite substantial for tort law purposes.

For one thing, the costs of establishing causation are probably just as great under the tort law as for administrative agencies, only they are decentralized (because each plaintiff has to support his own case) and, as we have seen, causation is much more difficult to establish in torts. (One tort law litigator indicated that he had seen costs of $80,000 to $100,000 for a plaintiff to establish medical causation in asbestosis suits against defendants.[72]) In fact, since a number of independent tort suits might have to be brought as a result of a particular exposure, the costs of establishing causation for regulatory purposes may be less than the costs to establish causation for an equal number of victims in torts.

Once it is found that exposure to a toxic substance causes a serious disease, a firm would have to take preventive measures either to prevent future tort law suits or as a consequence of regulation, unless it decided that the costs of tort suits were less than those of taking preventive measures. Presumably, the preventive costs would be about the same under tort law or under sensible regulatory law, for it is not clear that a compassionate and just society should permit a firm merely to bear the costs of tort suits and ignore preventive measures.

Other Considerations for Using Administrative Remedies

There are some important additional advantages in using administrative agencies rather than the tort law to secure our rights. The regulatory law aims essentially at the prevention of harm or control of risks before they materialize (typically to large numbers of people) by indicating, often with a greater degree of specificity than the criminal or tort law, how certain activities must be done. "Sellers of canned food . . . must prepare and seal products in designated ways; electric utilities must build and operate nuclear power plants in conformity with a multitude of requirements; owners of stores and other buildings open to the public must clearly mark fire exits";[73] or firms or municipalities, too, are told in considerable detail how clean our drinking water, food, and air must be. The prospective preventive requirements of administrative law are also backed by civil and criminal penalties should firms or individuals not comply with them.

In prescribing specific harm-preventive behavior administrative law is in some ways more intrusive than the tort law and possibly more intrusive in some respects than the criminal law. Specific administrative regulations are typically part of a "comprehensive" scheme to achieve statutory objectives. It is also in some ways a fairly expensive aspect of the legal system because it has a permanent bureaucracy of both quasi-lawmakers and technical or scientific support staff. These administrative costs are borne whether or not harm occurs, unlike the tort law. However, it is an appropriate institution to use at least in principle in a number of circumstances because of inefficacies in other areas of the law or because of the importance of the interests at stake:

1. Where "risks are not apparent to potential victims,"[74] as is typical of exposure to toxic substances, administrative regulations have an advantage over privately initiated tort remedies. In such cases individuals cannot protect themselves or identify tortfeasors as might be possible were the risks more apparent.

2. Where administrative costs or costs of a permanent bureaucracy relative to the harms sought to be prevented are not too unreasonable,[75] this is a reason for considering the use of an administrative remedy.

3. Where the "dispersion of harm" over many victims would limit the "effectiveness of privately initiated approaches to the control of risk,"[76] tort remedies would not serve the victims well. There may be insufficient

incentives for any one victim to initiate legal action, and because of free rider
and coordination problems it may be difficult to secure cooperation between
victims. Thus, there is a reason to try to prevent the harms from arising in the
first place.

4. Administrative remedies have an advantage where, because of the severity of
   the injuries threatened by some activity (e.g., death or especially severe
   diseases) and perhaps because of the large numbers of people potentially
   threatened, it is important to try to prevent the harms from ever arising rather
   than trying to punish wrongdoers or compensate victims after the fact.[77]

5. Finally, administrative regulations may be superior (a) where a fair distribution of benefits and burdens associated with a business activity would require
   that those who benefit from imposing especially hazardous risks on others
   should bear the costs to ensure that such risks do not materialize and (b)
   where the use of regulatory law would achieve a fairer distribution of such
   benefits and burdens than would use of the tort law.

These general guidelines do not provide reasons for the regulatory law totally to
displace the tort law or other legal institutions such as the criminal law. One would
also want to retain tort and criminal law sanctions to provide backup protections for
our rights should the regulatory law fail in its protective function.[78] I return to this in
a moment.

If we apply these general guidelines to the case of workplace and environmental
health protections, we see there are good reasons for using administrative agencies
in addition to the tort and criminal law to protect our rights in these contexts.
Governmental agencies are likely to have better relevant information needed to
prevent harms than the victims of such harms, who may have especially poor
information, for example, about the occasion of harm and the responsible party.[79] If
firms might be judgment-proof against large (or numerous) tort suits, this is a good
reason for placing the costs of prevention on firms up front by means of administrative regulations. If such costs keep from the market businesses that cannot afford to
prevent the possible harms, then the health of the community may be better off, even
though the community will lose these firms' products. Environmental health harms
are especially likely to be "dispersed" over many victims, and we have seen several
difficulties in establishing responsibility for such harms. Both reasons argue for
using an ex ante approach by a public agency to control exposure to toxic substances.

Finally, although there may be substantial administrative costs to using the
regulatory law, because of the foregoing reasons and the severity of injuries threatened by exposure to toxic substances such as carcinogens, these costs in many cases
will not be unreasonable.[80] Of course, this is not to ignore the costs of administrative
bureaucracies or their regulations, for both are relevant in assessing the defensibility
of institutions. Moreover, the cost advantage for the tort law may not be as clear as
some claim. A number of writers have indicated that the "administrative costs" of
torts—lawyers' fees, investigative costs, court costs, can be quite large: only about
50% of awards go to claimants and as little as 15% may go to cover actual losses.[81]

The preceding discussion, merely sketching an argument for using the regulatory law to provide workplace and environmental health protections, suggests some of the main reasons for having administrative agencies control exposures to toxic substances in addition to the criminal and tort law. Obviously, more can be said on these matters, but the arguments sketched here provide the outline of a justification for using the different areas of the law to accomplish various social aims. One chooses an institution in part because of the social goals to be achieved, in part because of the strengths and weaknesses of the institution for the task at hand, and in part because alternative mechanisms for accomplishing the same aims are not as good. In the circumstances just described the use of administrative agencies appropriately designed may better provide workplace and environmental health protections than would the use of some other areas of the law. This is not an argument for displacing the tort law, for as we have seen it has an important, if not always reliable, backup function.

Yet administrative agencies suffer practical shortcomings that temper to some extent this perception of their theoretical efficacy. An agency is subject to great political pressures, which may slow or paralyze pursuit of its legislative goals. It may be "captured" by the industries it is charged with regulating. It is dependent upon legislative and administrative funding and can become so starved for money that it ceases to be effective. Civil and criminal penalties for violation of administrative regulations may be so minimal as to leave it without much enforcement power.

Political pressures have been particularly great on the environmental health agencies. Considerable lobbying influences the enabling legislation in Congress and in the president's office. Agencies are subject to substantial pressures when they are created and during the implementation of the legislation. Individual regulations issued by an agency are heavily lobbied, and then, of course, they are frequently litigated. Finally, and perhaps most surprisingly, the science that informs agency regulations can become highly politicized, which further tempers an agency's effectiveness.[82]

Gillette and Krier offer an additional argument for not shifting all toxic tort protections to administrative agencies. First, firms that produce risks tend to have much better access to agencies than does the general public. In addition, agency experts' definitions of risks of concern are much closer to those favored by risk producers than to the general public. Consequently, risks of concern *to the public* are likely to be underregulated by the agencies.[83]

Because agencies are so infected with politics that they may not effectively carry out their legislative mandate, there should be *no weakening of tort law* health protections. The tort law can serve as a certain kind of backup to the failures of administrative agencies. It cannot duplicate agency functions. And, to change the metaphor, if it serves as a "safety net" to protect the rights of those whom agencies fail, it is a net whose meshes are too large for some of the reasons we considered here and in Chapter 2. The tort law is not the optimal institution for "controlling" toxic substances. Nonetheless, this protective net can still catch the larger failures of administrative agencies—asbestos, DES, and paraquat, for example. And despite its weaknesses it may still provide considerable deterrence to the production of toxic risks.[84] However, the tort law can serve this backup function only if the scientific standards for establishing causation are not so demanding as to keep meritorious cases out of court as we have seen that Black's proposals might.

CONCLUSION

In principle there are several good reasons for preferring administrative agencies to the tort law to provide environmental health protections. However, these theoretical advantages should not blind us to some of the political shortcomings of relying on administrative agencies. For that reason such agencies should not replace the tort law, which can continue to serve a backup function to the agencies in our legal system. Both are needed. Administrative agencies' ability to provide protections from toxic substances is limited, however, unless their approach to risk assessment is improved. We turn next to that topic.

4

Scientific Procedures in Regulatory Agencies

Administrative agencies' risk assessment policies specify the scientific procedures used to predict harm or risks of harm to people from exposure to carcinogens. This chapter considers two major risk assessment strategies found in the agency policies—the ad hoc and the guideline approaches—as well as two recommended alternatives to them, and argues that both groups appear wrongheaded. To a greater or lesser extent these four approaches share several features that in the regulatory context can produce bad consequences. They all aspire to the ideals of research science to some substantial degree: to find more and better data about the substance in question, to withhold judgment until data and theories justify the conclusions, to use cautious inferences about causal mechanisms, to be skeptical toward conclusions, and to be cautious about adding to the stock of scientific knowledge by protecting against false positives.

One cannot quarrel with these goals. However, when they are used inappropriately or insensitively in administrative agencies, they can produce substantial bad consequences as well as good. Risk assessment strategies embodying these ideals tend to be slow, to be insensitive, to be costly for agencies and society as a whole, and to frustrate the health-protective goals of environmental health legislation. A better approach would address uncertainties by policy choices (already done to some extent), determine the minimum amount of information needed to guide decisions and then use it, develop rebuttable interim standards for protecting public health, find an appropriate balance between false positives and false negatives for the context, expedite the evaluation of known carcinogens, and increase the rate at which substances are identified as toxins.

Here as in the tort law we face a paradigm choice in how we think about evidentiary procedures. We should avoid the temptation to demand more detailed scientific evaluation on a case-by-case basis of the substances considered for regulation. Giving in to this temptation will only exacerbate already serious problems. The regulatory challenge is to use presently available, expedited, approximation methods that are nearly as "accurate" as current risk assessment procedures, but ones which are much faster so that a larger universe of substances can be evaluated. I indicate two such procedures that appear to satisfy these conditions. The scientific challenge is to refine existing procedures and to develop others to expedite the identification and assessment of carcinogens.

Finally, since public policy in this area is developed partly by legislative mandate, partly by administrative policies, and partly by judicial review of those laws and policies, courts should allow agencies the discretion to utilize such methods. Recent court decisions suggest that they would defer to agencies in such practices.

The first section reviews some major laws aimed at protecting human health from environmental toxins in order to indicate some of the legislative strategies for protecting the public. The following section evaluates shortcomings of several extant and recommended approaches to risk assessment. Solutions to some of the problems that plague risk assessment are then suggested, and court regulatory decisions are reviewed.

INSTITUTIONAL BACKGROUND

Major Laws Regulating Carcinogens[1]

In the early 1970s, the U.S. Congress passed a number of laws aimed at protecting human health and the environment from exposure to toxic substances and other pollutants. These indicate some of the legislative strategies for protecting the public from carcinogens. The concerns of these laws include the regulation of toxic substances in our food,[2] air,[3] streams and inland waterways,[4] workplaces,[5] and drinking water.[6] Two pieces of legislation were designed to regulate more generally the pervasive presence of toxic substances in our lives.[7] In addition, one was written to prevent health and environmental problems from arising from abandoned toxic waste sites by requiring their cleanup.[8]

By one accounting, 21 different laws may be used to regulate carcinogens (Appendix D). Most of the statutes do not single out carcinogens for specific consideration but merely regulate them as a species of toxic substances. A few, however, have provisions aimed directly at carcinogens; one, the Food, Drug and Cosmetic Act, has special statutory provisions for regulating carcinogens as distinguished from other toxic substances, while several others—the Clean Water Act, the Toxic Substances Control Act, and the Resource Conservation and Recovery Act—mention carcinogens specifically.

Premarket Regulatory Statutes

Some statutes require premarket review or approval of a substance before it can enter into commerce. This requirement is seen in parts of the Food, Drug and Cosmetic Act (FDCA), and the Toxic Substances Control Act (TSCA), and in the Federal Insecticide, Fungicide and Rodenticide Act (FIFRA). However, there are some substantial differences between the kinds of premarket approval in these statutes. Both FDCA and FIFRA require a firm seeking to market a new substance to demonstrate its safety before it can go into commerce. The FDCA's well-known Delaney Clause, named after U.S. Representative James Delaney of New York, whose hearings led to the amendment,[9] provides that no intentional food additive

> shall be deemed to be safe if it is found to induce cancer when ingested by man or animal, or if it is found, after tests which are appropriate for the evaluation of the safety of food additive, to induce cancer in man or animal.[10]

Thus, if appropriate evidence indicates that a food additive is carcinogenic, the FDA may not consider it safe and must prohibit its use in food.[11] The manufacturer has the burden of proving that food additives are safe before they receive approval to enter the market. Provisions for color additives[12] are similar to those for food additives; that is, carcinogenic color additives are to be prohibited by FDA.

Under FIFRA an applicant for registration of a pesticide must file with the EPA certain required information, including a statement of all claims made for the pesticide, directions for its use, a description of tests made upon it, and the test results used to support claims made for the substance.[13] In a typical registration procedure a prospective registrant, usually the pesticide manufacturer, submits an application for a registration. If complete, the registration package's toxicity studies are evaluated, and if appropriate, the agency sets food safety tolerances required under Sections 408 and 409 of the Food, Drug and Cosmetic Act. FIFRA requires that the EPA ''shall register'' a pesticide if its composition warrants the proposed claims for it and its labeling and other required materials comply with the requirements of the act.[14] The agency may refuse to register a pesticide after giving the applicant notification of this intention and opportunity to correct the deficiencies in the application.

The Toxic Substances Control Act has a third premarket review approach. In general, anyone who intends to manufacture a ''new'' chemical must notify the EPA of his or her intention 90 days before manufacture is to begin. The company must submit a ''premanufacture notice'' (PMN) which contains information about chemical identity, proposed uses of the chemical, the expected production volumes of the chemical for the various uses, expected by-products, estimates of the numbers of people likely to be exposed in manufacture of the chemical, and methods for disposal.[15] The PMN must also include information on any toxicity testing that the company has performed, although the TSCA does not require that any testing be done prior to submission of a PMN.

The EPA has 90 days to review the PMN, although this period may be extended for an additional 90 days. The EPA's review can result in any of four actions: (1) the substance may be manufactured without restriction; (2) the substance may be manufactured for uses described on the PMN, but the agency can require that it be notified if any significant new use is considered; (3) the manufacture, processing, use, distribution, or disposal of the new substance may be regulated pending the development of additional information about it; or (4) the manufacture, and so on, of it may be regulated because it presents or will present an unreasonable risk.[16]

A major difference between the TSCA review procedures and the FDCA and FIFRA approval procedures is that under the TSCA it is as if the substance submitted for approval is placed on a regulatory conveyer belt and unless the EPA takes it off for closer scrutiny, it is permitted to go into commerce. Under both the FDCA and FIFRA, the agencies are under no time limitations to review the substance in question.

Postmarket Regulatory Statutes

A much larger number of statutes, including parts of the FIFRA and TSCA and the other environmental health statutes, provide for postmarket regulation of substances after they have been in commerce and people have been exposed to them. Such laws might require an agency to find that there is a health problem and then propose a regulation based on that finding, as in the Clean Air Act, the Clean Water Act, the Safe Drinking Water Act, and the Occupational Safety and Health Act. Still other laws might require an agency to find that there is a health problem, establish this fact in court, and seek some judicial remedy on that basis. Some sections of FDCA require this for foods contaminated by naturally occurring environmental carcinogens.

One state statute is worth mentioning in this context. California's citizen-passed initiative, the Safe Drinking Water Act of 1986 (the so-called Proposition 65),[17] introduces a substantially different legal mechanism. Under this law, industry has the burden of proving that exposures to or discharges of carcinogenic (or reproductive) toxins pose no significant risk to human health. Although not required to do so, the state has issued "no-significant-risk levels" to aid businesses and facilitate implementation of the law. Industry or environmental groups can challenge the scientific basis of no-significant-risk levels, but the burden of proof remains on the regulated community to show that some other level in fact poses no significant risk.

Agency assessment of the health risks under a premarket or postmarket regulatory arrangement combined with the regulatory science procedures for that agency will have substantial implications for the effectiveness of regulation. There are four main possibilities. An agency may operate under a premarket or a postmarket statute. In addition, either the agency may have the burden of proof to establish that a substance poses a risk to health or the manufacturer (or registrant or owner) may have the burden to establish that it does not pose a risk.[18] These yield the possibilities seen in Table 4–1.

If an agency must show a risk of harm from a toxic substance under a postmarket statute, then the more difficult it is to establish the scientific claims, the more substances stay in commerce even though they might pose risks of harm. The easier it is to establish the causal claims, the easier it is to regulate the substances and, other things being equal, the fewer harmful ones stay in commerce. If the manufacturer must show safety before introducing substances into commerce, then the more difficult it is to establish the scientific claims, the fewer substances enter commerce. If, as under California's Proposition 65, the manufacturer has the burden to establish that exposures to a substance do not cause a significant risk of harm to humans, this

Table 4–1 Regulatory Mechanisms and Burdens of Proof

Burden of Proof	Statute in Effect	
	Premarket Regulatory Statute	Postmarket Regulatory Statute
Agency must show substance poses risk	No statutes requiring this	Numerous statutes require this, e.g., SDWA, CWA, CAA, OSHA
Manufacturers/emitters must show substance does not pose risk	FIFRA, FDCA (Delaney Clause), TSCA	California's Safe Drinking Water Act of 1986 (Proposition 65)

provides incentives for the firm to ensure the safety of its products and to expedite the production of appropriate scientific evidence to this end. Thus, even for substances in the marketplace, this procedure expedites their evaluation and, if they are harmful, their removal from the market. Clearly legislation such as Proposition 65 has incentives that expedite consideration of the health effects of substances, but there are few such laws.

Procedural Requirements

Except for the few parts of statutes that require court-ordered remedies, most agencies authorized to regulate toxic substances under federal law must follow procedures mandated by the Administrative Procedure Act or similar procedures. In regulating substances an agency must follow these procedures for agency "rulemaking" in order to "issue a rule." Such rules may be issued according to rulemaking procedures that range from the relatively informal to the formal, resembling proceedings in a court of law. In general, the agency must announce in the *Federal Register* that it is proposing to regulate a substance (or group of substances), and it must describe the nature of the proposed regulation.[19] The agency must also give interested parties "an opportunity to participate in the rulemaking through submission of written data, views or arguments"[20] Following the comment period the agency usually holds hearings during which interested parties may have their comments heard. After considering both written and oral comments, the agency issues a final rule. This rule has the force of law unless it is challenged in court and invalidated for some reason by the appellate court.

Apart from these common features, informal and formal rulemaking are distinguished by the nature of evidence presented during the notice and comment period and at the hearing itself, as well as the standard of judicial review of agency action. Generally, under formal rulemaking an agency must conduct quasi-judicial proceedings with the opportunity to cross-examine witnesses.[21] Agency decisions following such proceedings are in theory more closely scrutinized by the courts if the regulatory decisions are appealed.[22] Under most of the statutes considered here, the agencies act under the requirements of informal rulemaking.

We return to some of these procedural matters later, because they hold the key to the limits of agency discretion in developing new scientific strategies to evaluate toxic substances. The main idea is that these procedural requirements plus the substantive statutory requirements and assignment of burdens of proof establish the legal hurdles an agency faces in controlling toxic substances.

Substantive Statutory Requirements

There are differences between the substantive requirements of the statutes. That is, Congress normally requires that regulations issued under each statute measure up to some substantive standard of health protection. Different laws reveal different attitudes toward risk. Some statutes reflect attitudes quite averse to human health risks posed by chemical substances. The most extreme example is the Delaney Clause of the FDCA considered earlier. The risk to human health under that statute is the only factor taken into account; no other social considerations outweigh such risks. This is a "no-risk" statute.

Other risk-based statutes use different statutory language. The Clean Air Act and Clean Water Act both make risks to human health the primary factor by setting the goal of regulating with an "ample margin of safety."[23] The Safe Drinking Water Act requires that enforceable regulations for drinking water be set with an "adequate margin of safety."[24]

Another approach is risk–risk balancing: weighing the risk to human health from exposure to a regulated substance against the risk to human health from not having the substance in commerce. The Food, Drug and Cosmetic Act appears to permit this kind of risk–risk balancing for food additives approved by the Food and Drug Administration prior to 1958.[25] For human drugs, FDA uses a "risk–benefit" approach, although again the primary factor involves the benefits and risks to patient health.

Some statutes are "technology-based" laws. These may require, for example, the agency to reduce emissions from a particular source to the extent this may be achieved by technological devices placed on the emitting source. Some such statutes require the "best practical technology" or the "best available technology." "Such regulations do not force new technology, but bring all control efforts up to standards established by existing control technologies."[26] Other technology-based statutes might be "technology forcing" because "new techniques may be required to achieve" some predetermined level of pollutant concentration.[27]

Still other statutes permit agencies to balance the risks to human health from carcinogens against benefits to be obtained by consumers, manufacturers, and others by permitting the substance to be in commerce. This is a risk–benefit balancing statute. Congress used the term "unreasonable risk" in the TSCA and FIFRA to refer to this kind of balancing.

These substantive requirements of the statutes implicitly establish the *distribution* of benefits and burdens under the statutes, that is, who bears the risks of uncertainty and what considerations if any outweigh potential risks from toxic substances. Thus, for example, risk–risk balancing statutes compare only the risks from permitting an exposure to a substance with the risks of not permitting such exposures. Such laws would not allow minor benefits to many people, which could total millions of dollars, to outweigh severe losses to a few who might contract cancer as a risk–benefit balancing statute might. The substantive requirements of the statutes distribute benefits and harms in much the same way that moral theories of distribution would. The procedures governing regulation also distribute benefits and burdens. If the state must establish that a substance is harmful before regulating, then costs of uncertainty are borne by those who contract diseases in the meantime. If the manufacturers/emitters must show a substance is safe before exposing the public, then it, consumers of its products, and its shareholders bear the costs of uncertainty. What is important is the particular pattern of distribution resulting from substantive and procedural requirements. We return to these topics in Chapter 5.

APPROACHES TO RISK ASSESSMENT IN REGULATORY AGENCIES

A Brief History of Agency Use of Risk Assessment

Agencies operating under the statutes just described are not always required to do risk assessments on carcinogens under their authority;[28] however, most conduct them. Agencies began using risk assessment procedures to evaluate the toxicity of carcinogens around 1970. Before that time, risk assessment was generally limited to primitive techniques used by structural engineers estimating the probability of catastrophic events and FDA administrators setting standards for pesticide residues.[29]

In 1970 the Food and Drug Administration began considering rudimentary elements of risk assessment, specifying the number of species that needed to be tested to evaluate the carcinogenicity of a substance, the maximum tolerated dose, and a two-generation bioassay design. These procedures were then updated extensively in 1982. Later in the 1970s the FDA began using quantitative risk assessments for certain environmental contaminants found in food; in the 1980s these same techniques were applied to food and color additives.[30]

The EPA began using carcinogen risk assessments during regulatory proceedings on the suspension and cancellation of pesticides. And in 1976 the EPA established "a permanent organizational unit, the Carcinogen Assessment Group," to develop guidelines and to perform risk assessments.[31] In 1979 the EPA established its proposed risk assessment policies for airborne carcinogens and also developed procedures for evaluating the risks in water. In 1984 the EPA proposed a revision of its carcinogen risk assessment guidelines, which was finalized in 1986.[32] The Consumer Product Safety Commission published carcinogenic risk assessment guidelines in 1978, made them effective immediately, but had to withdraw them following litigation about the classification of a substance as a carcinogen according to the policy.[33]

The Occupational Safety and Health Administration has never proposed explicit risk assessment policies, although it developed guidelines for the assessment and regulation of carcinogens in 1977. In contrast to other agencies OSHA's view was that "quantitative risk assessment would be used only to set priorities,"[34] not to set regulatory exposure levels. In large part this is because of the nature of the Occupational Safety and Health Act (OSH Act). That act requires OSHA to set the standards to regulate hazardous substances which most "adequately assure" that to the extent feasible "no employee [exposed to a toxic substance] will suffer material impairment of health or functional capacity even if such employee has regular exposure to the hazard dealt with by such standard for the period of his working life."[35] Since the prevailing view among most experts was that there is no safe level for carcinogens, OSHA's approach, at least until the Supreme Court's "*Benzene* case,"[36] was to set the standards at the "lowest feasible level."[37] After that decision, OSHA changed its strategy (considered later).

A number of factors have led to the agencies' use of quantitative risk assessment techniques and their particular commitment to them. Part of the explanation is intellectual. Rachel Carson's *Silent Spring*, which brought attention to the destructive effects of pesticides, was called anecdotal and unscientific. This led to governmental attempts to study and quantify such risks.[38] The National Cancer Institute (NCI) subsequently awarded contracts for the study of the tumorigenicity of 120 chemical substances.[39] This was further supported in 1971 by substantial congressional funding of the NCI.[40] Once such studies were begun, there were reasons to try to make them more accurate and to use the evidence available or evidence that could be generated to determine the magnitude of risks and how much the community is willing to pay to reduce or eliminate them. An editorial for the Society of Risk Analysis captures this attractive feature of risk assessment:

> [The Society for Risk Analysis's] thrust is the sane and systematic application of *knowledge* to environmental problems. SRA is now the model throughout the world for establishing rational tools to properly manage limited environmental resources: it is a symbol of sanity—showing the world how to do things better![41]

The notion of "accuracy" may be somewhat misplaced in this context. There are unavoidable and substantial uncertainties in the knowledge of the biological mechanisms involved, and statistical studies are in many cases too insensitive to provide the most accurate data. Thus, it is difficult to know what the true risk is from a risk assessment. And even at its best, risk assessment is a mixture of scientific and policy judgments. Nonetheless, there is a temptation on the part of most scientists and in the agencies to adopt for purposes of risk assessment the procedures typical of research science.

Additional motivations to use risk assessment procedures were provided by appellate court decisions. The 1980 Supreme Court's *Benzene* decision probably increased agency use of quantitative risk assessment. A plurality of the Court held that before issuing occupational health standards, the Occupational Safety and Health Administration must show that employees face a "significant health risk" at existing levels of exposure. Dicta in the case suggest that OSHA must demonstrate that significant health benefits likely will occur at the reduced exposure level set by a revised standard of exposure. And the agency must prove these conditions by a preponderance of the evidence.[42] Although a number of commentators believe this is a strained or wrong interpretation of the Occupational Safety and Health Act,[43] it nonetheless stands as a major precedent for OSHA's standard setting and may have had considerable influence on agency use of risk assessment procedures.

Both before the *Benzene* decision and after, OSHA claimed not to perform risk assessments.[44] Now, however, agency personnel acknowledge the use of such procedures in identifying "significant risks" to worker health; indeed, since the *Benzene* decision "OSHA is required to perform quantitative risk assessments to determine whether occupational exposure to toxic substances puts workers at significant risk of material impairment of health."[45] For evaluating substances the agency chooses "the best methodology based on the available evidence for a particular substance."[46] In short, it individualizes a risk assessment for each substance.

Commentators disagree about the influence of the *Benzene* decision on agency behavior. Some see the Court in that case as imposing "a difficult threshold barrier to rulemaking that has resulted in underregulation."[47] Others think that subsequent Supreme Court and circuit court decisions have been quite deferential to OSHA and other agencies.[48] We return to this issue later.

Although the *Benzene* case is the most visible legal case influencing the use of quantified risk assessments, a second decision from the Fifth Circuit Court of Appeals (which first ruled against OSHA on the benzene regulations) involving the Consumer Product Safety Commission may also have been important. In 1983 in *Gulf South Insulation v. Consumer Product Safety Commission* the court ruled that regulations banning the use of urea–formaldehyde foam insulation were invalid, since there was not substantial evidence in the record necessary to support the commission's action. The court found problems both with the human studies used as a basis of regulation, and with the animal studies.[49] Its most striking conclusion was that "it is not good science to rely on a single experiment, particularly one involving only 240 subjects, to make precise estimates of cancer risk."[50] And it had earlier remarked that "in a study as small as this one the margin of error is inherently large. For example, had 20 fewer rats, or 20 more, developed carcinomas, the risk predicted by Global 79 [CPSC's computerized risk assessment program] would be altered drastically."[51]

The court, however, not only misunderstood the nature of animal-based cancer risk assessments, condemning them as "not good science," but it failed to realize the animal study was based on a reasonably large study of animals.[52] Furthermore, its reason for rejecting the study is like saying in a murder case that if the victim had not died, the defendant would not have been guilty of murder. In addition, the court seemed to complain that exposures to which the animals were subjected were higher than typical human exposures,[53] again a standard procedure in animal bioassay studies.

This decision by the Fifth Circuit Court of Appeals, as well as its earlier *Benzene* case decision,[54] indicated considerable hostility to governmental regulations and may have forced agencies to support their cases with better quantified data before review in this circuit. Since an agency cannot be sure in which court of appeals its regulations will be reviewed, it must ensure that its decisions will survive review in the least deferential court. This may also have pushed agencies to use quantified risk assessment.

Apart from these two high-visibility cases, court scrutiny of regulations may have contributed to use of risk assessments. Commentators, surveying court reviews of agency regulations, indicate courts have had substantial general effects on agency behavior. One author studying all EPA litigation found that the federal courts "have had a significant effect on the policies and administration of the EPA."[55] "Compliance with court orders has become the agency's top priority, at times overtaking congressional mandates."[56] This led to budget "reprogramming," reduction in the discretion and autonomy of EPA administrators, increased power of the legal staff, and decreased power and authority of scientists.[57] It also increased the power of certain program offices within the EPA, increased their resources as well as staff motivation and morale, and lifted administrative burdens and prolonged review from the Office of Management and Budget.[58] When a decision is in favor of the EPA this increases its external power, authority, and discretion as a whole.[59] Nonetheless, response to court action is not "the best way to formulate environmental policy or to set our nation's environmental priorities."[60]

The findings reported in the previous paragraph indicate that courts have been active in reviewing agency actions and that this may have motivated the EPA to adopt quantified risk assessments. Such review makes the agencies sensitive to court concerns and likely to modify their behavior as a result. Furthermore, if a favorable court decision "increases an agency's external power, authority and discretion as a whole,"[61] an unfavorable one likely will have contrary negative effects, decreasing its external credibility, power, and authority, and in the extreme, possibly threatening its survival. Agency court losses may be demoralizing and may have chilling effects on future regulation, because time, hard work, and resources will have gone into efforts which then fail.

In addition, since setting standards is a highly visible and important political act, and since the agency "necessarily considers, as a matter of political survival, the often conflicting concerns of all interested parties," the agency may tend "to move as slowly and cautiously as possible."[62] One cautious approach in such circumstances is to use risk assessment methods to do the most careful assessments possible in the circumstances. Such procedures have the appearance of neutrality and impartiality and the appearance of following the procedures of research science, thus minimizing controversy and helping to ensure the agency's influence, authority, and discretion.

Moreover, courts may have reinforced agencies' adoption of the procedures similar to those used in research science: to acquire more and better data about the substance in question; to devote extensive efforts to ruling out alternative hypotheses; and especially to devote substantial resources to documenting that alternative, but frequently quite implausible approaches have been considered. Some standards of court review, such as the substantial evidence test (discussed later), may require more of this than others.

However, there are less demanding review standards for many statutes and even the degree of scrutiny under the substantial evidence test can vary considerably. Thus, although courts have reinforced the use of risk assessment and the particular way it has been used in the agencies, they can also influence the form such justifications take in the future. We return to this point shortly.

An additional influence that led to detailed risk assessments was the Office of Management and Budget (OMB) during the Reagan and Bush administrations. OMB, complying with presidential orders, demanded detailed justification of regulations, including a cost and benefit analysis showing that government resources were used more efficiently, and in many cases "actively opposed and delayed adoption of OSHA's standards" as well as those of other agencies.[63] Both the required justifications and OMB's opposition may have contributed to agency use of quantified risk assessments (QRA). An economic cost–benefit analysis of regulations is quite difficult without the use of some form of risk assessment.

Furthermore, most regulatory decisions are litigated.[64] In response to the litigation OSHA has found that "if we use generally accepted methodology, the courts will uphold our risk assessments. For us to have our risk assessments upheld in court, we cannot go too far out on a limb in any direction. Therefore, we tend to use mainstream risk assessment methodology."[65] This certainly suggests that OSHA personnel are conservative actors and thus committed to widely accepted QRA methodology and cautious in its use. Other risk assessors are likely similar to OSHA's in this regard. Of course, because of the *Benzene* case OSHA's actions may have had and continue to have higher visibility than those of other agencies.

A larger point from this revelation is that at present "mainstream" risk assessment procedures must be those that have the most widespread acceptance. Currently, such procedures consist of case-by-case evaluations of each substance under consideration.

Thus, by now most of the U.S. federal agencies charged with environmental health protections have developed some kind of risk assessment policy for evaluating carcinogens.[66] The approach adopted by regulatory agencies, however, can have important social consequences. In particular it will affect whether a risk assessment procedure is more likely to produce false positives and overregulation or false negatives and underregulation, and thus whether it is more likely to frustrate or promote wider regulatory goals of the institution in which it is used. Thus, in the next section I discuss some strategies for risk assessment in order to clarify some of the normative and institutional issues that this quasi-scientific procedure raises.

Current Agency Risk Assessment Practices

Although Chapter 1 addressed the (quasi-) science of risk assessment, this section characterizes the procedures adopted in the agencies a bit further, in order to see how they affect regulation. It is difficult to have an accurate characterization of agency risk assessment practices, because there can be substantial differences between agencies and to some extent agencies may modify their practices depending upon the substance under consideration. In addition, an agency's stated *policies* about risk assessment which are easier to document than its practices may be different from the actual agency practices. Nonetheless, agencies appear to follow two principal strategies in risk assessment. On one an agency adapts its risk assessment practices to fit the data on the substance in question—we might think of this as a highly "individualized" risk assessment. This approach appears to be typical of the practices at OSHA and perhaps at the Consumer Product Safety Commission.[67] On the second strategy an agency has certain default procedures to guide choices under conditions of uncertainty, typically expressed in agency guidelines, which it follows unless sufficient evidence is offered for departing from a particular guideline in a particular case. Call this the "default" risk assessment. This approach appears more typical of the EPA, the FDA, and the California Environmental Protection Agency.

The "individualized" risk assessment is suggested by OSHA personnel, who describe their procedure as follows:

> The agency performs each risk assessment differently choosing the best methodology based on the evidence for a particular substance . . . [and it takes] a much more, if you will pardon the expression, "seat of the pants" approach to quantitative risk assessment, using the data we have to arrive at our best estimate of risk.[68]

The agency apparently has no antecedent policy commitment to particular models to resolve the many inference gaps described in Chapter 1.[69] It appears to make its choice of high-dose to low-dose extrapolation models, interspecies scaling factors, and so on, on a case-by-case basis, as seems appropriate for the substance under consideration. Thus, in each instance the agency must consider all the evidence about a particular substance and then decide which of numerous statistical and biological models for bridging the inference gaps is appropriate for that substance.

The second general strategy for calculating the potency of carcinogens is for agencies to adopt standardized "default" assumptions for many of the critical inference gaps that appear in risk assessment. These might be used to adopt a certain high-dose to low-dose extrapolation model, to adopt a certain interspecies scaling factor, to count benign tumors as evidence of carcinogenicity, and so on. Agencies using default assumptions choose one such model, the default model, partly on the basis of scientific generalizations and partly on normative risk management grounds. At present it is difficult if not impossible to validate these choices scientifically.[70] There is much about the use of default procedures that seems correct and I develop further the rationale for it later. However, both the "individualized" and present conventional default risk assessment practices have serious shortcomings. There are three generic critiques of present practices.

A radical critique of a somewhat caricatured view of risk assessment by Freedman and Zeisel argues that the present risk assessment procedures have "little scientific merit" and "not much by way of scientific foundation."[71] Furthermore, the multistage model, the high-dose to low-dose extrapolation model favored by most agencies, as "judged by ordinary scientific standards" cannot be justified.[72] Although these authors recognize the government need to make crucial health decisions in a "rough and ready way," they "see no evidence that regulatory modeling leads to better decisions than informal argument and find the latter more appealing because it brings the uncertainties into the open."[73] They recommend putting more resources into epidemiology or basic research on the causes of cancer and on the origins and magnitudes of species differences, rather than putting the resources into agency risk assessments. While many observers may acknowledge the lack of scientific support for a number of the risk assessment models, most would not accept Freedman and Zeisel's recommendation that risk assessment should be rejected altogether. There is a need to have some assessment of the risks from carcinogens and most commentators are not as skeptical as Freedman and Zeisel about the scientific plausibility of the models.

A second group of criticisms suggests that what is wrong with present risk assessment procedures is that they should incorporate more data, more complicated models, and more "accurate" science. The concern is that risk assessment models fail to take into account a number of facts that they should.[74] These critics argue that more accurate biological models should be used for high-dose to low-dose extrapolation and mouse to human extrapolations, that greater use should be made of pharmacokinetic models (models designed to estimate the actual levels of a toxic substance reaching target organs), that appropriate accounting should be given to threshold effects from exposure to toxic substances, and finally that, instead of estimating maximum plausible exposures to toxic substances, risk assessors should try to estimate as closely as possible the actual exposures to which people are subjected.[75] In short, risk assessments should be tailored to the substance in question and made more "accurate." This view appears to be held by a substantial portion of the scientific community (it is difficult to estimate how large) and by much of the regulated community. Since this is such an important objection it is considered at greater length in the next section.

A third critique (also discussed at greater length below) concerns the emphasis on the present level of accuracy in risk assessment as well as the slowness with which present risk assessments are done.

Even though carcinogen dose–response estimates, only part of the risk assessment process, are typically "relatively straightforward, the production of conventional assessments is typically time-consuming, taking from 0.5 to 5 person-years depending upon the compound."[76] By themselves, the default dose–response (or potency) evaluations take "roughly 2 months to perform."[77] However, potency estimates produced by means of typical agency procedures take longer. Some resources are spent on activities essential to the conventional analysis—"tracking down the pertinent literature, identifying the appropriate data sets for analysis, weeding out poor data, and determining whether or not the ancillary data on pharmacokinetics and mechanism of carcinogenesis warrant abandoning the generic risk assessment assumptions."[78] However, substantial additional resources are spent on non-critical activities in order to demonstrate that the assessor has given full consideration to all potentially relevant data. Many risk assessments, for example, contain detailed reviews of studies on pharmacokinetics and genotoxicity even when the data are obviously inadequate and do not impact the potency calculation. In addition, substantial resources may be devoted to producing non-default estimates for comparative purposes only and to document that a variety of alternative approaches have been considered.[79]

Thus, agencies spend considerable time preemptively preparing themselves and documenting reasons for not using certain information.[80] Even if an agency uses default procedures and minimizes its documentation to the essentials for the data sets and models utilized, this saves considerable resources but still takes about two months per compound. To evaluate 369 identified but unassessed carcinogens that are presently listed under California Proposition 65 would take "approximately 35 person-years."[81]

Even though agencies do not formally adopt something like the science-intensive approach to risk assessment discussed in the next section, they approach this in their present practices in order to protect themselves from criticism and potential litigation. They obtain or review information about the substance in question to the extent it is available, but they do not require even greater testing, which some advocates of science-intensive assessments would have them obtain. Nonetheless, their present efforts require extensive human and monetary resources and are time-consuming.

Two Unacceptable Recommended Approaches to Risk Assessment

One reaction to present agency practices is to recommend that risk assessment procedures should be based on better science, and that they should be more accurate, producing predictions of more "real" risks and fewer "theoretical" risks.[82] This view aspires to the ideals of research science: acquiring more and better data about the substance in question (rather than relying upon scientific generalizations), using cautious inferences about causal mechanisms, ruling out possible but remote alternative hypotheses, and being cautious about adding to the stock of scientific knowledge. There are two possible recommendations to achieve this: the more extreme complete and accurate science approach and the less radical science-intensive approach.

The Complete and Accurate Science Approach

This approach advocates using only the most complete and accurate science in order to arrive at estimates of risks of harm to human beings from toxic substances. On this view, if peer-reviewed scientific information does not provide a *complete* understanding of the disease-causing mechanisms for a careful evaluation so that an accurate evaluation of the risks is possible, there should be no regulation.[83] This approach appears to resemble the one held by Bert Black discussed earlier concerning the use of scientific evidence for purposes of tort law compensation.[84] Few or none may actually hold this view, since it would require a *complete* scientific understanding of a substance and the risks it poses before it can be regulated, but it is one polar position.

The effect of such a procedure depends upon the statutory authority under which an agency operates. If it operates under a *postmarket* regulatory statute, which requires the agency to show risks from substances already *in* the market, this view would prevent almost all risk assessments and thus almost all regulation. If, by contrast, an agency operated under the authority of a *premarket* regulatory statute, which typically requires some kind of governmental *approval* and finding of *safety* of a substance before it can enter commerce, then few or no substances would be permitted to enter the market because there is insufficient understanding of them. If assessments could not be done, there would be no release of the substances into the market. However, as we saw earlier, most statutes in the United States regulating toxic substances are postmarket regulatory statutes.[85]

We can imagine a polar extreme to this view, that is, a deemphasis of detailed science (or almost no use of it at all) in risk assessment, with emphasis almost solely on policy considerations or qualitative risk judgments.[86] The most extreme of such views would be that science should play no role in the regulation of toxic substances. No one appears to hold this view, although some greatly deemphasize the role of science in regulation.[87] I do not consider this position here.

The Science-Intensive Approach

Although apparently no one holds the previous view, a number of writers hold positions that approach it, urging a science-intensive view, which emphasizes the most careful science applied intensively to each substance up for review, be used in risk assessment. Published work, anecdotal evidence, and personal observation suggest that support for this view is widespread in the scientific and in the regulated communities. This approach merits consideration because it reveals some of the shortcomings of being too committed to a particular view of science in the regulatory context.

Seilkin, Barnard, and Anderson to varying degrees appear to hold such views.[88] I call this the science-intensive view, for although its advocates urge the use of the most accurate currently available science for risk assessment, they appear not to demand that agencies have a *complete* scientific understanding of the mechanisms of a substance and the harm it causes before it can be regulated.[89] I also sometimes refer to this as the "conventional procedure," for although it departs somewhat from current agency practices in demanding even more scientific evidence than they typically do, and is recommended as an improvement on them, it is similar enough to them to consider them together.

In determining carcinogenic potency advocates of this view urge that agencies rely on the best possible, the most thorough, and the most adequate quantitative scientific input into the regulatory process that can be generated. Agencies should, they suggest, obtain more information about substances than they presently do. Risk assessors should

  test for the biologically effective dose in animals (not the administered or the target dose);
  rely upon more accurate high-dose to low-dose extrapolation models;
  use more realistic potency measures rather than conservative, health- protective, 95% upper confidence estimates;
  use both upper 95% bounds and lower 95% bounds when making high-dose to low-dose extrapolations;
  rely upon clinically indicated effects such as discomfort, disability, or death for evidence of harm to animals or human beings rather than the mere early stages of tumor creation;
  report as nearly as is possible the realistic risk from exposure to toxic substances;
  display not only a number indicating the risks to human beings from exposure to toxic substances, but also the range of uncertainty associated with such numbers; and
  estimate as closely as possible the actual exposures to which people are subjected.[90]

This view has several attractions. First, it appears more nearly biologically correct. Second, even when it is not entirely correct, if uncertainty ranges are assigned, risk managers and courts have a more accurate picture of what is known and what is not, as well as the degree of certainty of each belief. Accurate information *about the substance under evaluation* is the prized virtue on this view.

In addition, its advocates claim that since possibly 3.5–6% of the U.S. national income goes to controlling risks,[91] these costs should be reasonably related to the true magnitude of the risks created by toxic substances. Money spent on controlling unrealistic risks of harm to human beings might be much better spent on other things people want or on saving lives in other ways, for example, through reducing infant mortality, providing better prenatal care for low social economic class mothers, or providing better health care delivery to those who can least afford it. Thus, with more accurate information we will be better off ''in terms of both human health and economic well-being.''[92] As persuasive as these recommendations may appear, they nonetheless present several problems.

Attempts at accuracy, mistakes and false negatives. No one opposes correct scientific information, at least in contexts in which truth is the primary aim, but the attempt to obtain as nearly accurate scientific information as is possible for purposes of regulation has both good and bad consequences, which should be evaluated in the design of risk assessment and regulatory programs. For the most part the problem is the false negatives as well as the slowness and costs which the time and effort spent carefully evaluating each substance induces.

(Before addressing some of the issues of false negatives, we should note a caveat about the phrase ''attempt to be accurate.'' While advocates of the intensive science view emphasize this, there are respects in which they may ignore issues about which there should be greater accuracy. For example, the human population is quite varied and presumably sensitive to different levels of toxic substances. Infants, old people, those already suffering from disease may be much more sensitive to toxic substances than a 70-kilogram working male in good health. An accurate approach to risk assessment would take such reactions into account.[93])

Sometimes the attempt to be accurate can pose problems of its own. For example, the attempt to have more fine-grained knowledge about the risks from a particular substance may actually increase the number of mistakes agencies make and may add to the misconceptions associated with risk assessment. The use of pharmacokinetic information in risk assessment illustrates this general point.

Pharmacokinetics is the study of ''the concentration in target organs and the interaction of a biologically active agent with putative sites of action.''[94] The reason for using such procedures is to try to obtain more accurate information about the actual doses of toxic substances reaching target organs or tissues where the damage is done. However, in order to discover such doses, a good deal of additional information must be obtained: how the particular toxic substance in question is transformed and distributed in the body, the precise amount of the substance or its metabolites reaching the organs, and the organs in which it is likely to accumulate and which it may damage. For most substances such information is not available or is so poor that it cannot be used.[95] Where basic data are available, models have been developed to try to predict such events.[96] However, since this field is not well developed for evaluating toxic substances, there likely will be much that is unknown about the behavior of substances both in rodents, on which the tests will be performed, and in human beings, about whom the risk assessments will be made. Will use of such procedures reduce the uncertainties? On balance, will they lead to a better approach to risk assessment, all economic, social, moral, and scientific costs taken into account? Such questions are especially important when the risk assessment will be used in regulation and have legal force.

Underlying these concerns is a fundamental evidentiary point about the amount of complexity that can and should be incorporated into risk assessments. Catherine Elgin has argued that the more fine-grained one's evidentiary categories and the more evidentiary categories there are, the more chances one has for making mistakes for such refinements may actually "invite error and unreliability."[97] Fine-grained evidentiary categories require more careful classification of data under one category or another, and, if the classification is not quite correct,[98] one has made a mistake. For example, if we were asked merely to classify wines into reds or whites, nearly all of us would be infallible, although we might have some problems with so-called blush wines because of the fuzzy borderline. However, if we were given a more complex task, requiring the classification of wines into fruity, dry, and sweet whites and into similar subsets for reds, fewer of us would be correct. The more fine-grained such classifications become, the more detailed one's knowledge must be, the more opportunities there are for mistakes. This may also lead to disagreement and controversy, especially if one faces other experts in an adversary setting such as a regulatory proceeding or a court case.

What is appealing about fine-grained categorizations of knowledge is that if one "gets it right" and can defend it, one's knowledge is quite powerful. A wine expert's accurate classification of a wide range of wines into their proper categories is very impressive and much more impressive than my accurate classification of wines into reds and whites. Thus, whereas fewer and less fine-grained categories may lower the number of mistakes and thus increase accuracy, the accuracy is purchased at the price of the knowledge being in grosser categories. For some purposes this may be acceptable: for other purposes it may not.

Classifications serve various goals of the individual or institution that uses them. For some purposes simple rough-and-ready classification schemes work quite well—for example, for choosing the proper wine for a barbecue. For other purposes, a more subtle, fine-grained classification may be needed, say, for choosing wines for each course of a gourmet French dinner. There are advantages and disadvantages to different schemes depending upon the purposes to which they are put. Similar considerations apply to the evidentiary categories of risk assessment. That is, the usefulness of the potentially fine-grained knowledge one can obtain from pharmacokinetic studies must be evaluated in terms of the goals it serves, the cost of obtaining the information, the benefits in accuracy (if any) this may permit, and the opportunity costs of performing such assessments and ignoring other substances.

A National Academy of Sciences panel has provided reasons for using pharmacokinetic information:

> Regulatory agencies are to make assumptions that they believe to be conservative, so as to avoid underestimation of the risk to the human population. Unfortunately, overestimating risk might unnecessarily *eliminate jobs* or *commercially important materials,* thereby *decreasing our general standard of living.* Hence, the more we can replace empirical assumptions with experimentally validated procedures, the better off we will be in terms of both health and economic well-being.[99]

This account holds out the benefits of more detailed experimental information. A major motivation behind such concerns appears to mimic the goals of research science: obtaining data peculiar to the substance in question (rather than relying on less substantiated generalizations), having more experimentally validated data rather than less about a substance, using cautious inferences about causal mechanisms, and using caution in adding to the stock of scientific knowledge. This account also emphasizes normative, nonscientific reasons for using pharmacokinetic information: lowering the cost of regulation and the impact on the economy. Despite the scientific motivation and the normative considerations, however, there can be costs to attempts at precision, especially for a field in its infancy.

For one thing, even though the use of pharmacokinetics promises greater accuracy, such information and accuracy are not available for most substances. Detailed information would have to be provided on a substance by substance basis, and more data would be needed to validate a model *for each substance*. Until we know much more, it seems that the operative assumption would be that each substance metabolizes, distributes, accumulates, and harms organs in unique ways—few generalizations at this time are permissible. Discovering such mechanisms requires detailed data and careful modeling. Each substance must have a unique analysis. This takes time and resources.

Further, if Elgin's arguments offered above are correct, the more fine-grained one's analysis is through a detailed biological study that must be chemical- and animal-specific, the more sources of mistakes there are. Mistakes generate uncertainty about whether one has the correct answer. In a regulatory and adversary setting this will only add to the controversies. The more sources of controversy there are, the more time it will take to discuss and to settle the issues and to rule out the possibility of mistakes or alternative explanations. The more tests and careful analyses one tries to do in order to predict risks to human beings, the greater the dollar, time, and human costs are. It will take more money, put highly trained and talented Ph.D.s to performing what may be fairly routine and boring tasks,[100] lengthen the time it takes to perform risk assessments, increase the size and complexity of risk assessment documents, and invite additional preemptive documentation. The National Academy of Sciences summarizes some of the concerns:

> Clearly, the more detailed the model, the greater will be the theoretical utility, but also the more numerous will be the sources of uncertainty and the data necessary to estimate the values of the parameters of the models. In practice a compromise between complexity and simplicity must be reached.[101]

Simpler, less complicated, and somewhat more rough-grained models for estimating risks may generate fewer mistakes (although models would be somewhat less precise and might produce larger mistakes when they occur), elicit fewer sources of controversy, be less time-consuming, and result in somewhat less complex risk assessment documents. It is difficult in the abstract to indicate the proper balance for the various social concerns. In some circumstances where data are readily available and where pharmocokinetic models are well developed and the results can be quickly generated, such information may be more accurate and reduce costs. Where these conditions are not satisfied, we should use more approximate numbers and omit detailed studies.

However, what is clear is that *the value of more detailed information is determined by normative considerations appropriate to the institution or the activity in question*. The kind and amount of information one needs and gathers is in part a function of the uses to which the information will be put. Norms, in short, determine epistemological and evidentiary considerations. We return to this later.

Thus, even if there is a gain in certainty by attempting more detailed scientific inquiries, at what price is marginally more accurate information worth the effort? Research suggests that pharmacokinetic models will typically make a difference of a factor of about 5–10 in the ultimate risk number, although in extreme cases the results can be greater.[102] Is such a difference in the ultimate outcome worth the cost, effort, and possible chances of additional errors from a field in its infancy? (In rare circumstances a fivefold difference might justify the additional costs of detailed testing if good data were available and enormous costs were associated with such a difference.)[103]

More seriously, are these costs worth it when such information is unavailable for most of the 300–400 identified carcinogens and the 50,000–100,000 substances in the market which have not been properly evaluated for their toxicity? I think not, but only a more careful analysis of the costs of detailed studies and of the respective costs of false positives and false negatives, as well as the costs of overregulation and underregulation, would settle such questions. An aspect of the answer to this question is considered later.

This discussion of pharmacokinetics also serves as an example or an analogy for other detailed investigations into the properties of carcinogens for regulatory purposes. On the science intensive view additional information for pharmacokinetics and other aspects of risk assessments likely will have to be generated on a *chemical-by-chemical basis*. Providing the appropriate data and finding the appropriate biological models for the substance in question will necessitate detailed case-by-case evaluations. This in turn is likely to generate further discussion and controversy. The more details that go into a risk assessment document and into a regulation, the more agency effort is needed, the greater the amount of documentation needed about choices made, and the more opportunities the losing side in a regulatory struggle has for objecting to the regulation, whether that side is the affected industry or those concerned to protect human health. This effort might produce somewhat more accurate information. However, this is both difficult to ascertain (because of uncertainties) and to the extent it can be realized does not seem a great improvement over faster approximation methods which are very close to conventional risk assessments. And, it might result in mistakes, which is especially problematic in an adversarial setting.

More importantly, the opportunity costs may be great, because the effort and resources devoted to preventing false positives and overregulation and to pursuing accuracy for each substance considered may result in an agency not testing for other harmful substances, thus producing false negatives, or in an agency not evaluating identified toxins, thus producing regulatory false negatives or underregulation. Procedures designed to minimize risk assessment and regulatory false positives and overregulation may therefore result in false negatives and underregulation. We return to this point later.

Some shortcomings of uncertainty ranges. A second recommended feature of the accurate science view is that nearly all scientists advocate including uncertainty ranges as part of the numerical estimates of risks that are provided in risk assessment documents; this is a constituent feature of scientific practice and, as we saw in

Chapter 1, epidemiologists are suggesting this as an improvement on statistical tests of significance. Some commentators even suggest showing alternative models that could have been used at each step in the risk assessment process. Use of uncertainty ranges provides a better picture of what is and is not known about the substance under consideration. Nonetheless, the use of uncertainty ranges, while appropriate, even essential in research science, poses some concerns in regulation.

For one thing, at the present time it is not clear that risk managers will have the requisite appreciation of uncertainty ranges, thus possibly leading to misinterpretation or misunderstanding of risk assessments. Furthermore, the National Academy of Sciences argues against displaying all possible models at a choice point claiming that in the present state of risk assessment this suggestion is confusing, does not avoid the problem of having to choose between different models, and may lead to more ad hoc policy judgments.[104]

An equally important concern is that regulatory agencies by providing uncertainty ranges may invite more intrusive court intervention and invalidation of regulations. For example, in *Gulf South Insulation v. Consumer Product Safety Commission* the Fifth Circuit Court of Appeals, apparently not understanding the nature of animal bioassays and seemingly concerned about the uncertainties involved, seriously misunderstood both the nature of the evidence, which ordinarily would be considered quite good, and the amount of uncertainty likely to be present in regulatory science.[105] Displaying the wide uncertainty ranges that could arise from using different assumptions might well occasion greater court intrusion into the process. Such intervention may occur, even though the agencies using the best data and theories available are being as conscientious as they can be and are making as responsible a set of risk assessment and risk management judgments as possible. If explicit uncertainty ranges were used, courts would have to be educated to be more sensitive and more discriminating in their reading of regulatory documents and might have to change their standards of judicial review in order to be dissuaded from invalidating regulations simply because they contain great uncertainty ranges.

An additional problem is that the assignment of wide uncertainty ranges may also be inconsistent in many cases with the present evidentiary requirements of the law.[106] How serious a problem the assignment of uncertainty ranges poses depends upon the sophistication of the courts as well as their deference toward agency rulemaking. Court deference to agencies is considered later.

The social costs of science-intensive risk assessments. A third major concern typically advanced by those holding an intensive science view is more obviously normative:[107] agencies may produce costly regulations by regulating too stringently because they have not accurately identified the ''real risk'' but instead are regulating nonexistent or minimal risks. This can lead to inefficient use of scarce resources. For example, if people spend too much to pay for mistaken or too stringent regulation of substances (through the cost of goods that they purchase), they may have substantial opportunity costs. In particular, overestimating risks may ''eliminate jobs or commercially important materials'' or reduce our standard of living.[108] This is a nonoptimal allocation of resources. Moreover, if we do not have accurate predictions of risks, then we may be spending money in a misguided attempt to save lives that are not endangered or spending too much money per life saved. Thus, we are saving fewer lives than we could with the same amount of money. As the Academy of Sciences indicated, the more accurate the assessments of risks are, "the better off we will be in terms of both health and economic well-being."[109]

These arguments explicitly bring wider risk management and regulatory concerns and policies to bear on the seemingly factual approach to risk assessment. I believe this cannot be avoided, but such concerns must be assessed as management or regulatory strategies.

The foregoing arguments express economists' standard concerns about the efficient use of resources. The concern is that if markets are working efficiently, then resources are "wasted" on regulating risks that are apparent but not real, or on regulating real risks too stringently compared with the economic benefits to be gained from regulation and with opportunity costs. Science-intensive risk assessments may reduce false positives and overregulation, thus saving resources associated with the substances under review. (Of course, the most accurate assessments should also aim to reduce false negatives, but we have seen why the attempt to achieve accuracy may frustrate this aim.)

Before directly evaluating these normative considerations, we should note that a major presupposition of this argument—that there is a "real risk" easily discoverable in the circumstances—is not obviously correct. In principle there are "objective" risks from carcinogens. However, in the present state of knowledge, they are not easily discoverable and the public health costs of delaying action until these are known with scientific certainty may be unacceptable.[110]

The first argument above correctly assumes that resources not spent for one purpose will be spent or saved somewhere else, and it expresses a legitimate concern that regulations can be too expensive. But the concern with expensive regulations focuses on overregulation and regulatory false positives for the substance under review.

Although this can be a problem, efforts to prevent overregulation of a few chemical substances in the present have substantial opportunity costs, for they may produce too little regulation of substances not yet considered and thus incur false negatives. We can see this point in detail by considering the costs of ignoring identified but unevaluated carcinogens because of the focus on careful evaluation of each substance. Consider the universe of 369 chemicals listed under California's Proposition 65.[111] As of April 1991 detailed risk assessments had been performed on only about 74 of the 369 carcinogens.

Conventional procedures, which evaluate each substance considered for regulation about as well as present scientific information permits, have left hundreds of identified carcinogens unassessed, which may well impose health costs on the public. One can assess the opportunity costs of conventional methods by modeling the total social costs of conventional risk assessment and faster alternatives to them. These are summarized in Figure 4–1. To account for the possibility of errors the simulation compares the costs of "mistakes" of expedited and conventional risk assessment procedures. The notion of a "mistake" is a technical term; it is not that assessors make calculation errors, or that they are careless. Instead, because of uncertainties, poor data, and poor understanding of mechanisms in performing any

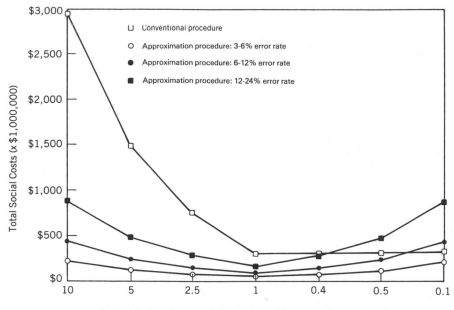

Ratio of Cost of Underregulation to Cost of Overregulation per Chemical

Figure 4–1   Total social costs of evaluating 369 Proposition 65 carcinogens.

risk assessment risk assessors may not discover the true risk (from a God's eye view) of a substance. Conventional risk assessments might result in mistakes by being slow and leaving identified carcinogens unassessed as well as mistakes from uncertainties, poor data, and so on; expedited procedures might make mistakes by not assessing known carcinogens quite as intensively from a scientific point of view as on conventional procedures. In modeling social costs values are assigned for the costs of *individual* assessment mistakes of a certain kind—false positives, false negatives, overregulation and underregulation—and then *aggregated, summed,* according to the kinds of mistakes the two different kinds of procedures might make.

For a universe of 369 substances conventional risk assessment procedures are compared with three hypothetical expedited procedures with different "error" rates for major overregulation and underregulation and minor overregulation and underregulation. Out of 369 identified carcinogens only 74 have been evaluated by conventional methods. The model assumes that conventional methods result in 6% of the substances that are regulated being overregulated in a major way with 94% assessed as "accurately" as present procedures provide. This leaves 295 or 80% of the 369 carcinogens unevaluated. These are the equivalent of regulatory false negatives, because they are unregulated substances.

For approximation procedures the model assumes that the total universe of identified carcinogens can be evaluated quickly and subsequently regulated because the procedure is much faster. However, because the expedited methods are approximations to conventional risk assessments, these may result in departures from conventional results and from the true risks. Thus, the model assumes several different error rates for the approximations are possible. The low–error rate scenario assumes that 3% are underregulated in a major way and 3% are overregulated in a major way. *Major* overregulation and underregulation means that the potency values derived from the approximation procedure departed from the conventional method by more than a factor of 25. Risk numbers that are off by a factor of 25 constitute a major divergence sufficient to expect different kinds of regulatory action. There typically is no bright line between the kinds of regulatory action until the differences between possible risk assessments are large; a factor of 25 is such a difference. For instance, the control strategies for a risk of 10^{-5} might be quite different from control strategies for a risk of 25×10^{-5}. Moreover, since there legitimately can be a range of outcomes that are all reasonable from the same data, risk management considerations are likely to guide the regulatory strategy, unless the risk numbers differ too much.

The low–error rate scenario assumes that 6% of the substances are overregulated in minor ways and 6% are underregulated in minor ways. *Minor* overregulation and underregulation means the potency values differed from conventional potency values by a factor of 5–25. Approximation results are assumed to be "accurate" if they differ from conventional procedures by less than a factor of 5.[112] Higher error rates were assumed for two other approximation scenarios.

The idea behind this simulation is that misregulation produces social costs: carcinogens regulated too little for the harm they cause impose health and other costs on the public; carcinogens that are overregulated for the harm caused impose costs on industry, their shareholders, and the public. Economists commonly assume that regulatory false negatives cost $10,000,000 and regulatory false positives $1,000,000.[113] Cost numbers are difficult to arrive at for modeling purposes and open to debate. On one hand, there are substances that are carcinogens, but which pose no risk to human health because they are no longer in use or they are industrial intermediates in a closed manufacturing process with no human exposure. On the other hand, economists have found on the basis of labor market studies that the dollar value of an *individual* life lost varies from $1.8 million to $9.2 million and the value of a life lost calculated on the basis of union wage studies varies from $3.2 to $21.[114] Authors find a best estimate from both kinds of studies of about $4–8 million for a single life.[115] However, this is the estimated cost we are willing to pay to prevent the premature loss of *one life,* not the costs in terms of human life and health effects that *one unregulated substance* would cause. Moreover, while some carcinogens may pose no risk because there is no exposure, others, most notably asbestos, may cause death and disease for many. And, of course, some substances will be more valuable commercially than others and there will be some variance in the costs of overregulation and regulatory false positives. Rather than enter that discussion, I begin with the traditional economists' assumption that regulatory false negatives, major underregulation, and minor underregulation are ten times as costly as their counterparts (regulatory false positives, major overregulation, and minor over-regulation, respectively). This ratio is then modified to 5:1, to 2.5:1, to 1:1, to 1:2.5, down to 1:10 to model how the total social costs associated with the mistakes resulting from each kind of procedure would change as a consequence. The results are summarized in Figure 4 1.

By failing to regulate 295 identified carcinogens the conventional science-intensive procedure is much more costly than any of the three hypothesized approximation alternatives. The cost advantage continues for all approximation alternatives (including the least "accurate") until the costs of underregulation are assumed to be 40% (1:2.5) of the costs of overregulation for the approximation scenario with the highest "error" rate. [The other approximation scenarios (with lower "error" rates) continued to be economically preferable to even lower cost ratios.] This cost assumption, however, appears contrary to the literature and to common views about the relative importance of underregulation to overregulation, which in general regard the health costs of underregulation and false negatives as greater than overregulation and false positives.

The pattern of "nested" approximation curves will extend vertically up the graph as the approximation methods have greater "error" rates. However, it appears that approximations have a cost advantage over conventional procedures for reasonable ratios of the costs of regulatory mistakes and for reasonable error rates for approximation methods. (The experience of California in using a particular risk assessment procedure indicated that the hypothetical low–error rate approximation procedure is quite realistic (discussed later).)[116]

These results illustrate two points: (1) There are substantial social costs to present risk assessment procedures and the more "scientifically based" alternatives to them because they are slow—they delay consideration and evaluation of identified carcinogens. (2) Approximation procedures, which more quickly evaluate the whole universe of known carcinogens, result in much lower overall social costs. The largest social costs are opportunity costs—not evaluating known toxins quickly enough. Thus, it appears better to evaluate the universe of known carcinogens even if there are some mistakes than to assess a smaller group more carefully while ignoring a large number of identified carcinogens. There is a second important philosophical question concerning costs: On whom should the risk of regulators' mistakes fall? Such questions cannot be answered in the abstract, but some of these issues are addressed in Chapter 5. However, a focus only on the normative costs of avoiding false positives and overregulation, the main emphasis of the science-intensive approach to risk assessment, ignores other important considerations.

Thus, to return to the main argument, a presupposition of the "efficient use of resources" argument that slower, more science-intensive assessments reduce social costs is not obviously correct. Although this approach in some cases may reduce costs for the substance under consideration, when opportunity costs are considered these savings appear to be swamped.

An additional problem with the efficient use of resources argument is that economists who hold this view often subscribe to a normative moral theory for the allocation of resources that most philosophers find problematic, the philosophy of utilitarianism.[117] This foundation of the efficiency argument is so controversial that most philosophers and some economists reject it as an unacceptable theory for distributing benefits and burdens in a community.[118] I return to this topic in Chapter 5.

The second version of the efficient uses of resources argument assumes that resources saved from the accurate regulation of toxic substances can be used to save even more lives through other programs. As morally persuasive as this argument seems, it is not free from difficulties. For one thing, it assumes as a matter of institutional or market fact that any resources not spent to save lives in one area will or can be spent to save lives in another. Although in principle this *could* be done, it seems unrealistic to assume that *markets will* obviously and efficiently allocate resources from one life-saving context (e.g., regulation of toxic substances) to another (e.g., decreasing infant mortality). Markets are not stratified into life-saving and non–life-saving marketplaces.

A variation of the argument assumes that *governmental institutions* would reallocate funds from one life-saving context to another. This seems plausible at least in principle and especially for administrative agencies under a single executive such as the president. Moreover, in recent years the Office of Management and Budget under the Office of the President has tried to do this. However, to hold this view, one needs considerable optimism about governmental efficiency and wisdom in the face of institutions that are largely independent of one another (e.g., EPA and FDA)[119] and of institutions that have different congressional and public constituencies in support of their activities.

These factual assumptions aside, however, the underlying principle seems attractive: one should use resources efficiently in saving lives, and if more lives can be saved by attacking natural causes of death such as malaria rather than man-made causes such as synthetic toxic substances, then funds should be allocated accordingly.[120]

Even this principle, however, is not without controversy, although it cannot be developed in full detail here. One presupposition appears to be that the loss of life from one cause is just as important as loss of life from another cause. A variation of it is that *risk* of loss of life from one cause is just as important as a numerically identical risk of loss of life from another cause. Both are controversial and I would argue indefensible on moral grounds. They also appear not to be supported on empirical grounds, given how people think about the issues.

The argument is not free from controversy because, for example, it is one thing to die in a natural disaster, such as an exploding volcano or an earthquake, and quite another to be a victim of a murder or of a reckless or negligent release of a toxic substance. In each case the victim is dead. However, human agency and human fault makes a difference in our judgments of the issues. Morally faulty human actions are more blameworthy and frequently of greater cause for concern than acts of nature. It is, however, difficult to give general guidance here because the issues are quite complex. Clearly a natural plague such as the Black Death is of great concern because of the harm it could do. A similar plague caused by human beings would be of equal concern for preventive purposes, but we would also want to know whether they were culpable in causing it.

Moreover, is saving lives from natural causes morally as important as preventing human beings from inflicting preventable harms on one another, everything else being equal? The former rests on a principle of beneficence concerned with the prevention of suffering, whereas the latter rests on one or another principle of justice.[121] Justice principles are concerned with *distributing* benefits and burdens (distributive justice), with *correcting* wrongfully inflicted harms and wrongful gains at the expense of others (corrective justice), or with *punishing* wrongdoers and *preventing* humans from harming one another (retributive justice). Which is appropriate in the circumstances depends upon the issues involved. The issues are complex, not easily answered, and controversial, contrary to the presuppositions relied upon for the normative claims in support of the science-intensive approach. The important point, however, is that the normative considerations invoked for the science-intensive view are not obviously correct and they are controversial.

Moreover, the *acceptability* of *risks* to life from different sources have a different normative status. It is one thing to incur risks to one's life where the risks are palpable, where one fully understands what one is getting into, and where one voluntarily takes or embraces them. It is quite another to be put at risk by the knowing or unwitting activities of others when those activities could have been done more safely. Thus, for example, it is one thing for a person deliberately to study an exploding volcano with full awareness of the risks to her welfare, and quite another to be exposed to risks to her health by having to drink water polluted by the deliberate or negligent or even inadvertent disposal of toxic substances into the nearby groundwater.

In addition, empirical studies show that in fact the public explicitly makes such distinctions regarding the acceptability of risks. Paul Slovic has shown, for example, that "riskiness means more to people than 'expected number of fatalities.'" He notes that lay people's

> basic conceptualization of risks is much richer than that of the experts and reflects legitimate concerns which are typically omitted from expert risk assessments. As a result risk communication and risk management efforts are destined to fail unless they are structured as a two way process.[122]

In short, the public appears to be quite discriminating in judging the acceptability of risks of the same magnitude. A *normative* evaluation of the acceptability of risks suggests similar results.[123] Thus, a number of legitimate factors influence judgments of the acceptability of risk other than the numerical odds of being harmed in similar ways.

An additional presupposition implicit in the arguments for accurate risk assessments is that one always has a duty to prevent as many deaths as possible. Stated in this bald way, the claim is indefensible.[124] A somewhat weaker consideration in the regulatory context is whether a third party, a governmental regulatory agency, should use its authority and resources to intervene to prevent a firm from possibly harming a small number of people as a consequence of its activities (because they are exposed to a substance that causes cancer, for example) or use its authority and the same amount of money to prevent the deaths of a larger number of people from natural causes. And weaker yet is the question of whether a government facing a choice between saving more people (from death from natural causes) and saving fewer people (from death from human causes), where neither alternative would violate anyone's rights, should save the larger number. The answer to this may appear obvious, but it rests on a view that has not been without controversy in moral philosophy. Such issues merit more extensive discussion than can be provided here.[125]

It is not necessary, however, to explore these issues in detail in order to pursue the argument of this section. The normative considerations that appear to favor the accurate science approach and seem obvious are not; they are controversial. The broader social consequences of using different risk assessment strategies are pertinent to evaluating their desirability. In fact, taking into account human resources and total social costs in an attempt to make risk assessments more accurate seems to argue against such an approach.

The science-intensive approach to risk assessment may invite more mistakes than it initially appears and frustrate some risk assessment and regulatory goals because of the focus on preventing false positives and overregulation. It also begs some of the *normative* issues at stake in the debates about risk assessment and regulatory strategies, and it may not rest on the most defensible *normative* strategy (although I merely suggested this view). By placing primary weight on avoiding regulatory false positives and overregulation, it appears not to give much (or any) weight to avoiding false negatives and underregulation over the whole universe of chemicals, which in the regulatory context may be more important. And we have seen how costly this can be on traditional economists' assumptions. Thus, it appears less than fully defensible. Nonetheless, I have focused on this view, for it is one to which many are tempted and attracted.

SHORTCOMINGS OF PRESENT AND RECOMMENDED PRACTICES

Several shortcomings common to present agency practices and the recommended accurate science alternatives to them emerge from this survey. The concern to make risk assessment more scientifically well founded is understandable, yet it seems to me that this approach is wrongheaded in the current state of knowledge.

First, the concern to find accurate ways to reduce uncertainties suggests an attitude of scientific caution that attaches to the assessment of the risks of the substances in such circumstances. To the extent that scientific caution urges risk assessors to withhold judgment while obtaining more and better information or making the models more accurate, many of the undesirable consequences discussed earlier will emerge. Many uncertainties could be resolved by a policy choice rather than attempting to achieve greater biological accuracy, which would be difficult to achieve, expensive to produce, and time-consuming, even if it were possible. Some uncertainty cannot at present be eliminated.

Second, a desire to have an accurate evaluation of the risks of each substance on a case by case basis may move risk assessors to insist on a degree of accuracy in standards of evidence for risk assessment that is typical of ordinary scientific inquiry. Although this concern is laudable for a variety of purposes, such evidentiary standards have to be examined very carefully and used only with great wisdom and care for regulatory purposes. As we saw in Chapter 1, it is possible for the demanding evidentiary standards typical of research science to prevent discovery of precisely the kinds of harms that should be the object of regulation. Consequently, in the pursuit of regulatory goals, risk assessors should be wary of unthinkingly and unwisely subscribing to such standards when investigating phenomena for regulatory purposes.

Third, present priorities focus primarily on avoiding false positives and overregulation for the regulatory process as a whole. We should distinguish between the underregulation and overregulation effects of risk assessment procedures for each substance considered from the underregulation and overregulation effects of the regulatory process for all substances. *For each substance considered,* current practices appear to strike an underregulation and overregulation balance by utilizing health-protective assumptions in choice of inference guidelines and by trying to be as accurate as possible in evaluations. However, *for the process as a whole* present procedures tend to ignore the universe of known but unevaluated carcinogenic substances; this results in substantial opportunity costs because of underregulation and regulatory false negatives. Present practices also appear to give insufficient attention to the universe of 50,000–100,000 of chemicals most of which have not been identified as toxins or not.

Fourth, present agency practices and the science-intensive alternative both are slow. This is in part because of the concern for science-intensive evaluation and caution. (They may also be slow because of a concern to develop "perfect" regulations.[126]) The rate of *research* science is typically not a relevant factor guiding the research (apart from competition between researchers or in unusual cases of research on dread diseases such as polio or AIDS, for example). However, failure to be concerned about the rate of identification of carcinogens and the rate of the assessment and regulation of risks from known carcinogens perforce results in false negatives and underregulation. We incur hidden social costs as a result.

There is a more worrisome message in the statistics about the rate of risk assessments. As we saw, examination of the false positive and false negative rates for conventional and hypothetical approximation procedures reveals the organized social forces favoring the conventional method. If a faster procedure were adopted, there might well be more mistakes than at present (although this is not clear). Some of these will affect the regulated community, but it has substantial ability to lobby the agencies and to limit this result. The large number of false negatives and underregulation on the conventional procedure may affect a group of unknown victims, but they typically have no organized lobbying power to limit mistakes affecting them. They might not even know the causes of their diseases and they are unlikely to communicate with one another or to lobby the agencies. Thus, a number of political forces favor the present procedures and perhaps favor an even slower process. This does not imply existing practices are desirable.

The specific criticisms just summarized are a species of a more general criticism: present and recommended practices fail to take into consideration some of the major social consequences of the practices in question.[127] In fact they appear to underemphasize social and public health consequences that may be more important for the goals of regulatory institutions than the ones that motivate present practices; they appear to ignore concerns about avoiding false negatives and underregulation and concerns to expedite risk assessment and regulatory procedures.

This criticism of present regulatory science practices (and the recommended accurate science alternatives to them) echoes the findings in Chapter 2: our epistemology, our standards of evidence for causation, should be appropriate for the area of human endeavor in which they are to be used. For preventive health purposes the present practices, as well as the recommended accurate science alternatives, should be suited to serve better the institutions in which they are used.

AN ALTERNATIVE APPROACH

The general strategy of this book is to uncover and take seriously many of the major social consequences of risk assessment and regulatory procedures. Given the difficulties of present agency practices and recommended alternatives to them, what is an appropriate approach to risk assessment in the agencies in order to implement the legislation under which they act and to protect the public from toxic substances? In answering this question I treat risk assessment as a regulatory tool, useful for achieving certain aims—a position not greatly different from that of some industry representatives.[128]

Agencies can acknowledge explicitly, if they have not already, the number of policy considerations implicit in risk assessment. And they should rely upon them more than they do. Substantial uncertainties can be addressed by policy choices.

Next, agencies should scrutinize the data requirements and the inferences used in risk assessment and risk management decisions to see whether demanding standards typical of research science procedures are likely to frustrate the goals of regulation. Appropriate standards of evidence must be chosen relative to the context in which they will be used.

Further, they should take into account the rate at which risk assessments are done. Other things being equal, agencies should use faster rather than slower risk assessment procedures. Even faster procedures that produce some overregulation and underregulation have social cost advantages over present practices.

Finally, however, the solution to improved regulatory use of scientific information is not solely in the hands of regulatory agencies. Since overall results are also influenced by court decisions concerning the initial legislation as well as agency interpretation and implementation of it, courts have a substantial role to play here as well. Implementation of these recommendations needs support by the courts, and it seems recent court decisions may provide appropriate precedents.

Coping with Scientific Uncertainty

Using policy choices to address uncertainty to some extent resembles present agency practice. However, agencies should more explicitly acknowledge the role of normative considerations in risk assessment and use them to reduce debate about which models are appropriate for the substances under consideration for regulation. As the National Academy of Sciences notes, "risk assessment must always include policy, as well as science."[129]

This proposal has two parts: (1) Policy considerations should address the number and amount of uncertainties in risk assessments by guiding the choices of models used. Agencies should be even more committed to this approach than they are at present. (2) This further commitment involves engaging in less debate about application of inference guidelines to particular substances; the procedures followed should be applied somewhat more rigidly. Agencies should also engage in less preemptive research and less documentation of procedures not followed.

Explicit incorporation of policies can help prevent certain kinds of mistakes and shape risk assessment and risk management by appropriate institutional considerations. Default inference guidelines to address uncertainty and to choose between models are similar to burdens of proof in the law—they predispose the agency to a particular decision where there is an inference gap, and they are to be followed unless there is a substantial amount of credible evidence offered by the regulated party for adopting different inference guidelines. And the amount of evidence needed to depart from the default should be set high enough so that agency personnel do not spend all their time hearing appeals to the chosen default guidelines. We will consider two procedures for cancer potency assessments that illustrate these characteristics. Risk assessment is both too important and too uncertain to be left exclusively to the risk assessors,[130] so policy considerations should be explicitly endorsed by the agency, not left to individual risk assessors.

As we saw in Chapter 1, both the notion of "a risk" and the idea of "acceptable risk" are already normatively laden; this should be explicitly acknowledged and endorsed.[131] Risk assessment in the present state of knowledge is plagued by great uncertainties and probably cannot and should not be separated from risk management. (As scientific techniques and knowledge develop, this may be less true, and some studies can help reduce the extent of uncertainties.)[132]

Several reasons support the policy approach: (1) Given the inadequate scientific knowledge and data for risk estimates,[133] we have a choice in enforcing postmarket regulatory statutes in the face of uncertainty: either we do not regulate or we make decisions on the basis of available evidence and nonscientific policy considerations. Given the health-protective aims of most environmental health legislation, the choice seems clear. "Protective assumptions" and policy choices can guide decisions under conditions of uncertainty instead of waiting for costly, more accurate scientific information and a science-intensive evaluation of each substance. (2) As we have seen, there is inherent tension between the disciplinary norms of research science and good regulation.[134] Permitting the science perspective (without explicit adoption of policies) to dominate assessment can reduce public protection "against potential toxic hazards, increase regulatory decision-making costs, and expand opportunities for obstructive behavior by agency bureaucrats or private parties hostile to toxics regulation."[135] (3) Furthermore, "the illusion that risk assessment is a purely scientific activity reduces the visibility and *political accountability* of policy judgments that often guide regulatory decisions on toxic hazards."[136] In a democracy, making the policies more explicit after appropriate discussion better serves the regulatory aims of the legislature and the agencies. At present while agencies to some extent rely on policy considerations to guide risk assessment, they show some tendency toward more "biologically based" risk assessments, which as we have seen is problematic.[137] Such procedures should only be adopted if they promote rather than frustrate the aims of the enabling legislation and the health protection aims in particular.

In addition, the National Academy of Sciences suggests several reasons for using explicit policy considerations as inference guidelines: (a) It is better to use policies in the guidelines than to permit ad hoc decisions. Policy considerations will guide even ad hoc decision only they will not be explicit. Risk assessors making ad hoc decisions without publicly acknowledging policy guidance may inadvertently dominate the regulatory process. The goals and the policies in the authorizing legislation should be explicitly incorporated to guide decisions when the science runs out.[138] (b) The combined policy and scientific judgments that enter into the choice of guidelines should be made in accordance with the best scientific thinking on the topic and then reviewed periodically. (c) Explicit policy-guided inference guidelines promote fairness, consistency, and predictability for the firms affected and consistency between agencies. (d) Such guidelines provide a locus for debate for evolutionary improvement of the guidelines over time. (e) They may help public understanding. (f) And, most important, they can promote efficiency in the regulatory process—agencies can avoid rearguing every inference about which there is uncertainty.

Using policy choices to address uncertainty combines easily available scientific information with appropriate social policies when there is scientific uncertainty. Sources of social policy considerations which could guide risk assessments when the science is unavailable include (1) policies implicit in the enabling legislation such as in the OSH Act or the Safe Drinking Water Act;[139] (2) consideration of the cost effectiveness of analytical and scientific procedures used repeatedly on individual chemicals to ensure that the agencies are using efficient practices in regulation[140] and not incurring great opportunity costs; (3) consideration of the potential for catastrophic miscalculations that might result from widespread population exposures, absence of long historical records of human exposures, or evidence of unusual potency in a chemical substance;[141] (4) Protection against certain kinds of mistakes; and (5) broader moral considerations. The third and fourth reasons have to do with preventing certain mistakes; the first two seek to incorporate normative considerations of the legislation guiding the institution. The fifth reason acknowledges broader moral considerations we use to evaluate our institutions and guide interpretations of the law. And to the extent that guidelines are applied more rigidly, thus reducing the time for evaluating substances, this helps to expedite risk assessments, which reduces opportunity costs.

Howard Latin as well as Dale Hattis and John Smith have suggested how the normative aims of legislation might guide risk assessment.[142] Many of Latin's claims are negative. First, within EPA and OSHA there appeared to be no consistent pattern for choice of models to bridge the inference gaps. Sometimes models were chosen to avoid underestimating risks, sometimes middle-of-the-road choices were made, sometimes the models underestimated risks, and sometimes the treatments were ''methodologically convenient'' but relatively unrealistic—''they appeared without rationale.''[143] At OSHA, this appears to be deliberate policy because each assessment is unique.[144] Other agencies do not appear to be as ad hoc in their procedures, but, as Latin indicates, they may not fully succeed. Clearly agencies working under the authority of a particular statute should be more consistent. Elsewhere Latin has made specific recommendations about how burdens of proof for regulatory purposes should follow the organic legislation.[145]

Furthermore, frequently agency risk assessment policies appear not to have been influenced by the organic legislation authorizing agency action. Yet it would seem that an agency working under a highly health-protective statute such as the Safe Drinking Water Act, which requires that maximum contaminant levels for toxic substances be set with an "ample margin for safety" (which for carcinogens is zero), might well conduct risk assessments somewhat differently than an agency working under a statute such as the pesticide law (the Federal Insecticide, Fungicide and Rodenticide Act), which requires that the agency set residue levels so that they pose no "unreasonable risk to man or the environment taking costs into account." (Of course, if an agency is merely seeking to avoid certain kinds of mistakes, it may not refer as explicitly to statutory language.)

Hattis and Smith suggest how statutory policies might guide risk assessment in expressing scientific uncertainty in the risk assessment process:

> Under a risk/benefit balancing type of statute, the full probability density function for all sectors of the exposed population may be relevant to the decision maker's choice, whereas only an "upper confidence limit" . . . may be relevant under a statute that requires the decision maker to assure that the standard will "protect public health with an adequate margin of safety."[146]

The general idea: the choices among risk assessment models in part should be guided by the appropriate statutory authority under which regulators act, to the extent that statutes are clear enough for such purposes.

We can oversimplify and generalize a bit on the Hattis and Smith suggestion and suggest that a cost–benefit balancing statute bears some similarity to utilitarianism, which is considered in Chapter 5.[147] Their second suggested strategy, which aims to ensure "adequate protection for public health," is typical of the more protective environmental health statutes in the United States. This approach, which typically assigns greater urgency to avoiding false negatives and underregulation for distributive purposes, is more akin to that suggested by a theory of justice than a utilitarian theory. Justice theories might be thought to give a special place to health care protections, to give them priority over a number of other social goods. (These issues are considered in Chapter 5.)

In a democracy the policies that reduce uncertainty in risk assessments should be explicitly adopted in the quasi-democratic forum of agency hearings that are part of rulemaking. This way all participants, including the courts, know the policy orientation that risk assessments will have. Furthermore, the particular policy considerations guiding model choices need not be adopted forever; they can be revised as the relevant science dictates.

These recommendations suggest an important point about democracy. To the extent that risk assessment and the regulatory process more generally is seen merely as a scientific enterprise, public input is irrelevant, for the public is not expert on scientific issues. Once it is recognized that risk assessment (and regulation) is in part a function of policy considerations, public input, especially in a democratic form of government, becomes a relevant consideration to shape the process. Moreover, the recommendations reinforce a theme of this book: risk assessment and risk management strategies are also in part philosophical issues and the public has important input on them.

There is a second part to the recommendation of this section. Once agencies have adopted certain default models on policy grounds, they should be somewhat more "rigid" in their application of the default choices in order to expedite risk assessments. I later suggest two proposals that satisfy these conditions and that can improve the speed of conventional risk assessments. Of course, some data gathering is essential to track down pertinent literature, to identify appropriate data sets, to weed out poor data, and to determine whether there is good enough pharmacokinetic information on substances. (As we will see, even time spent on these activities can be minimized or eliminated by using data banks that already contain this information.) However, agencies should reduce preemptive research on alternative models and on data which are obviously irrelevant.[148] As part of this suggestion agencies should determine the minimum data needed for the risk assessment proposed (when combined by policy considerations) and when that is available proceed to evaluate the substance in question. Research and argument as conventionally done can take up to five person-years to produce a potency estimate for a carcinogen.[149] Clearly, this can be reduced. Moreover, agencies should tolerate any departures from default assumptions only if a high burden of proof is overcome. This would substantially expedite assessments, for less time would be spent debating particular inference guidelines. Of course risk assessments "individualized" for each substance such as OSHA employs will be quite slow and counter to the suggestion made here.[150]

Mitigating the Demanding Evidentiary Standards of Science

We have already seen that the demanding standards of evidence routinely used in scientific inquiries may unwittingly frustrate the discovery of substantial risks of harm even when they exist. Any facts produced by an epidemiological or other statistical study under less than ideal research conditions or even an animal study testing for a rare disease are already heavily infected by policy considerations. Scientists must render judgments of the very kind the substantive statutes require: How great a risk to human health should be countenanced? What risks should the study try to detect?

Moreover, assessors should be wary of "negative" studies; these are only a failure to find an effect, not necessarily evidence that there is no effect. We saw in Chapter 1 how easy it is for a study to be negative for reasons unrelated to the existence of harm and we saw in Chapter 2 some tort law courts have rejected studies for having sensitivity (power) that was too low to detect risks of concern. The power of such studies to detect a particular effect becomes quite important. Frequently, studies of rare diseases lack the power to detect low but substantial risks of harm.

In addition, agencies should be prepared to modify the use of the 95% rule, or of 95% confidence limits, if this will result in more sensitive interpretations of the data. In effect, risk assessors should be willing to regulate on the basis of clues that may not have the same degree of certainty or strength of evidence as traditionally demanded in research science. If regulatory agencies are to perform properly their preventive health protective function as required by virtually all environmental health laws, they must not wait until the demanding standards of research science have been satisfied, for as we have seen this can easily leave people at risk. Instead they should be willing to act on the basis of less stringent evidentiary standards and less than perfect evidence. Moreover, agencies should also resist the temptation to invoke unmeasured confounders to account for positive epidemiological studies, for this may also frustrate health protective goals of the agencies.

The foregoing suggestions, however, might increase the number of false positives, since confidence in positive findings may be reduced and there might be some mistakes. Thus, we must be prepared to tolerate some false positives as the cost of reducing false negatives; for regulatory agencies we must find an appropriate balance between the two kinds of mistakes. Adhering to the demanding standards of science gives great priority to protecting against false positives and deemphasizes or ignores the cost of false negatives. In the next section I consider some of the social benefits of finding a better balance between the number of false positives and false negatives, and in Chapter 5 I discuss briefly some of the theoretical underpinnings for this view.

Finally, a suggestion for using policy choices to mitigate the effects of both the uncertainty of risk assessment and the demanding standards of evidence that OSHA risk assessments might have on regulation is contained in Justice Marshall's dissenting opinion in the Supreme Court's *Benzene* decision concerning the regulation of workplace exposure to toxins. This adds to the preceding suggestions.[151]

Marshall, writing for a minority of four, dissented sharply from the plurality decision, which required OSHA to show a "significant risk" of harm at current exposure levels before regulation would be justified. He rejected the court plurality's "significant health risk" standard and the evidentiary requirements that attend it. In his view, when the "magnitude of risk cannot be quantified on the basis of current techniques," the Secretary should be permitted to act primarily on policy grounds.[152] Policy considerations could help reduce uncertainty, and agencies could act largely on the basis of expert opinion and policy judgments in some circumstances.

For one thing, the Secretary of Labor should act in accord with the remedial purpose of the Occupational Safety and Health Act,[153] which is highly averse to risks to workers' health (in effect showing greater concern for false negatives). If risks cannot be easily quantified by risk assessment procedures, they could be established by expert opinion,[154] akin to the practice of congressional committees or perhaps to tort law trials. Thus, clues to toxicity indicated by experts could be at least part of the basis of regulation. Second, an appropriate margin of error may be used to establish both risks and new exposure standards, subject to the "feasibility" requirement.[155] And policy guidance is especially appropriate for known carcinogens, since any "deficiency in knowledge relates to the *extent* of the benefits [that would be provided for exposed workers] rather than their *existence*."[156] Third, since the decisions "to take action in conditions of uncertainty bear little resemblance to the sort of empirically verifiable factual conclusions to which the substantial evidence test [of court review] is normally applied,"[157] courts should be more deferential toward agency actions. (We turn to this in the final section.)

The specter of unbridled agency discretion may have concerned the court in the *Benzene* case; the plurality in part sought to control what they viewed as agency overregulation. In light of a history of regulation which shows a large universe of substances unevaluated for carcinogenicity and a smaller number of identified but unassessed carcinogens, such concerns with overregulation should no longer loom so large. Agencies and courts should show greater concern with possible unidentified toxins and known but unassessed carcinogens. Both agencies and the courts should modify their policies to address these conditions.

In sum, I suggest that OSHA and other agencies use scientific techniques that are more rather than less sensitive to the risks of concern, be prepared to act on the basis of clues of harm and not wait for scientifically certain information, and use the same default assumptions for each substance, unless there is very good biological evidence for departing from them. These strategies will save considerable time and, it appears, not produce results greatly at odds with present procedures (I return to this point later). Finally, agencies and the courts should not demand such detailed risk assessments for substances under consideration that most of the universe of known toxins goes unevaluated or that we do not get to the task of identifying other toxins among the 50,000–100,000 known chemicals.

The aim in explicitly using policies to guide risk assessment and using less demanding standards of evidence is to avoid regulatory paralysis, to permit regulation where the most demanding standards to science might block it (but there is fairly good evidence available when supplemented by policies), and to expedite regulatory consideration of substances. The next section indicates some procedures that would greatly expedite the evaluation of known carcinogens.

Expediting Risk Assessment

The following paragraphs argue for implementing a complex recommendation that would achieve a better balance between regulatory false positives and false negatives than exists at present and which incorporates the policy orientation suggested above: agencies should develop approximation procedures to expedite the assessment of known carcinogens and use the results for interim regulations with binding legal authority. These approximation procedures explicitly incorporate the policy approach in risk assessment guidelines to reduce uncertainties and to reduce debate about choice of models.

There are several procedures that would expedite the assessment of risks from carcinogens for regulatory purposes. All of these are directed at obtaining reasonably accurate figures for the carcinogenic *potency* of substances, a process that can be quite slow.[158]

Both of the suggested procedures utilize readily available data from animal bioassays that greatly expedite potency assessments (which can take up to 5 years) and that effectively make the assessment of carcinogens nearly contemporaneous with the latest results from animal bioassays. Use of information in this data base facilitates several tasks that take considerable time in conventional assessments: performing the literature search, exporting data from the analysis, identifying the appropriate bioassay, and estimating the dose-response relationships.

Use of Tabulated TD₅₀s with the most Sensitive Sites and Species[159]

In the first procedure the so-called tumorigenic dose (TD_{50}) values from a data base are used to calculate the carcinogenic potency for human beings for the substance in question. The TD_{50} value is defined as the chronic dose (in milligrams per kilogram of body weight per day) that would produce tumors in half the animals that would have remained tumor free at zero dose over the standard lifespan of the species.[160] An agency could estimate the cancer potency from a data set by using the most sensitive target site in the most sensitive study known to researchers or, if there were several animal studies, by *averaging* the data from different studies or by using a data selection procedure mandated by the agency. This first procedure uses data from the most sensitive site and species.[161] Considerable time can be spent identifying the appropriate data set to be used as a basis of the potency calculation, but this task has been made easier. Gold and associates created the Carcinogenicity Potency Database containing the results of more than 4000 laboratory animal experiments on 1050 chemicals.[162] Use of these data reduces considerably research time for agency personnel because critical information on the data from animal bioassays is here.

In estimating the TD_{50} values, Gold et al. assume that the age-specific cancer incidence increases linearly with the daily dose to which a person is exposed.[163] When the background tumor incidence is small and the tumor incidence linear, the potency of the age-specific cancer is expressed in a simple mathematical relationship—the logarithm of 2 divided by the TD_{50} amount (the amount of a substance that causes tumors in 50% of experimental animals) (cancer potency = ln $2/TD_{50}$). Although this relationship may appear overly simple, it is derived from the standard high-dose to low-dose–response model, the so-called linearized multistage model. By assuming a negligible background cancer rate and a linear dose–response rate this simplified equation is easily derived. The derivation is provided in Appendix E. Finally, because the calculation is so simple, requiring only a calculator and some expertise to read the data base, it is extremely fast.

In addition, even for cases in which dose–response data are nonlinear and the background incidence is not negligible, the TD_{50} can be used to obtain relatively good approximations of the dose–response slope. Krewski and co-workers compared the slope of the straight line joining the TD_{50} value and the origin with the typical value produced by fitting the standard high-dose to low-dose model (the Crump multistage polynomial) to the same data for 585 experiments selected from the Gold et al. data base.[164] Potency estimates obtained by linear extrapolation from the TD_{50} were nearly always within a factor of 5–10 of those derived from the linearized multistage model. Had Krewski et al. used the formula just given instead of directly taking the slope of the line connecting the TD_{50} value to the origin, the ratio of the TD_{50}-based to multistage model–based slope would have been closer to unity. TD_{50} values tabulated by Gold et al. can therefore be used to estimate the potencies of carcinogens and these estimates closely approximate those derived from the conventional linear model used by most agencies.[165]

About 200 compounds out of 369 substances listed under California's Proposition 65 have adequate data for TD_{50}-based potency calculations; 74 of these have been the subject of conventional risk assessments at the time this first approximation procedure was evaluated. To illustrate how easily the process of producing cancer potency estimates for regulatory purposes can be accelerated, Cranor et al. have presented the 200 cancer potency estimates which result from applying this approximation method to the most sensitive bioassay results identified by the TD_{50} selection algorithm.[166]

To assess the accuracy of this accelerated method of estimating potencies, TD_{50}-based estimates are compared with a set of 77 (three substances have two potency assessments) conventional potency estimates derived by the California Environmental Protection Agency (CEPA), CEPA reviews of potencies derived by EPA, and additional EPA potencies currently under consideration for adoption in California.[167] This comparison is based on (1) conventional quantified risk assessments estimates, and (2) TD_{50}-based potency estimates derived from the most sensitive site in the most sensitive bioassay. The ratios of the most sensitive site and species TD_{50}-based estimates to conventional estimates is presented here.

The concordance between the most sensitive site and species TD_{50} and conventional results is surprisingly good, particularly considering the substantially different resources and time required by the two approaches. Figure 4–2 plots the frequency distribution of the ratio of the potencies derived from TD_{50}s for the most sensitive studies to potencies derived by CEPA and EPA using conventional procedures.

This degree of concordance is not too dissimilar from that which exists between conventional potency estimates produced by different regulatory agencies (compare Figure 1–7. The use of different default assumptions by EPA and CEPA (e.g., to correct for studies of short duration or with early mortality) results in comparable discrepancies between potency estimates for the same compound.[168]

TD_{50}-based potency estimates based on the most sensitive species differed from conventional potency estimates by more than a factor of 10 for 16 out of 77 comparisons (21%). Only 17% differed by more than a factor of 25. By taking the logarithm of these ratios the distribution can be further characterized: the geometric mean of the ratios of TD_{50} based to conventional estimates in 3, with one standard deviation corresponding to a factor of 6. That is, on average the TD_{50}s are no more than a factor of 3 different from conventional risk assessments. The differences in magnitude between potency estimates and factors that could potentially account for them are apparent. The main factors accounting for the discrepancies are nonlinearity in the dose–response data; apparent differences in carcinogenic potency in humans versus animals; and data set selection. Sensitive evaluation of the data sets, and application in some cases of a different model to the data sets can reduce the discrepancies between TD_{50} results and conventional potency asessments.[169]

For example, the TD_{50}-based method should not be used if there are adequate human data for risk assessment purposes. Also, the Gold et al. data base identifies bioassays with significantly nonlinear dose–response relationships, that is, bioassays for which the TD_{50} and other linear extrapolations are less likely to be accurate. Potencies derived for compounds with studies exhibiting substantial nonlinearity should be checked. However, the results presented in Cranor, et al. and the study of Krewski et al. indicate that significant misapproximations usually will not occur even as a result of nonlinearity in a specific data set. The data in the Gold et al. data

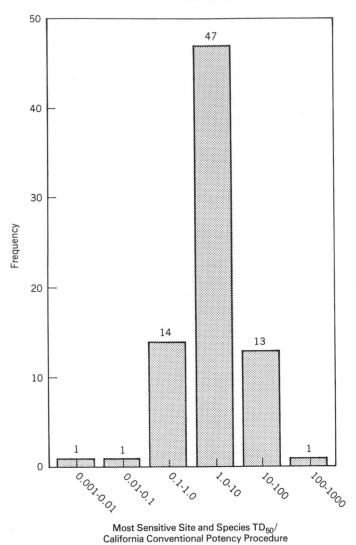

Figure 4–2 Ratio of potency estimates. [From C. Cranor, L. Zeise, W. Pease, L. Weiss, M. Hennig, and S. Hoover, "Improving the Regulation of Carcinogens: A Proposal to Expedite the Estimation of Cancer Potency" (forthcoming in *Risk Analysis*).]

base on any particular chemical can be scanned to find those cases where the TD_{50} algorithm disussed here failed to select more powerful studies performed at low doses which indicate substantial nonlinearity. Thus, information in the data base itself provides clues to substances for which departures from the policy-guided default assumptions are likely to be justified; additional attention can be devoted to these few substances, if necessary. The procedure as a whole consequently expedites the evaluation of the vast majority of substances under default procedures; for the few where the data justify more individual treatment, alternative assessments can be performed.

Furthermore, even if a substance is not in the data base, a literature search might well produce relevant data bases to which the expedited potency calculation could then be applied. The standardized data base saves research time, which can benefit agencies. But additional savings are realized by using an extreme default procedure—the TD_{50} procedure—for calculating potency.

Use of Expedited Linearized Multistage Default Procedures

A second procedure that will expedite the derivation of potency estimates for carcinogens uses the Gold et al. TD_{50} data here to select an appropriate data set chosen in accordance with agency data selection procedures, and then applies the standard linearized multistage (LMS) default dose–response model to that. The State of California has followed such a method for implementing a recommendation by its Proposition 65 Science Advisory Panel. That panel recommended the adoption of the TD_{50} carcinogen potency values, but both agency personnel and the public appear to be more comfortable using potencies derived from default data set selection as well as default extrapolation methodologies, since those are in accordance with standard practices and are legally authorized defaults. The LMS default procedure is nearly as rapid as the TD_{50} derivations because literature searches are restricted to the Gold et al. data base and no time is spent documenting procedures not adopted. In this case greater expertise is required to evaluate the data base and a computer is needed to perform the extrapolation. However, the procedure produces results closer to conventional assessments than the TD_{50} procedures described previously (since it utilizes the same default assumptions as the conventional method for the most part). Figure 4–3 compares the default expedited procedure with potencies derived from conventional methods.

The LMS default expedited procedure is even more "accurate" (closer to conventional potency assessments) than the TD_{50} most sensitive site and species procedure. For 78 substances evaluated by the LMS and conventional procedure by CEPA and Cranor et al., 2.6% were off by more than a factor of 25; only 9% were off by more than a factor of 10. The geometric mean (an average) is 1.24 and one standard deviation is 3.97 compared with conventional science-intensive procedures performed by CEPA.[170]

To complete the estimations of human cancer potencies from both the TD_{50} and LMS procedures, the potency values derived from animal data would then be multiplied by an appropriate interspecies scaling factor chosen on both scientific and policy grounds. This information can then be combined with appropriate exposure information to approximate more detailed conventional risk assessment.[171]

The Virtues of Expedited Procedures

The purpose of discussing alternative potency assessment procedures is not to recommend any one of them in particular (although the potency estimations based directly on TD_{50} values or fairly rigidly applied default procedures seem quite good), but rather to indicate the importance of expediting risk assessment and to indicate that plausible procedures are available.

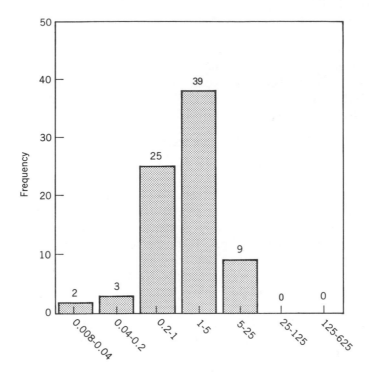

Ratio of Expedited LMS Default Potency/California Conventional Potency Assessment

Figure 4–3   Ratio of potency estimates. [From C. Cranor, L. Zeise, W. Pease, L. Weiss, M. Hennig, and S. Hoover, "Improving the Regulation of Carcinogens: A Proposal to Expedite the Estimation of Cancer Potency" (forthcoming in *Risk Analysis*).]

Both procedures are scientifically sound, serve the purposes of most environmental health statutes better than conventional risk assessments, better protect the public health, have lower social costs than conventional risk assessment, and save agencies money.

Both are scientifically sound. The TD_{50} procedure is derived from long-accepted standard default procedures for calculating potencies in conventional potency assessments, and the correlation between TD_{50} values and conventional California Environmental Protection Agency (CEPA) potency assessments is nearly as good as that between conventional risk assessments done by two different agencies such as CEPA and the Environmental Protection Agency. The expedited LMS default procedure (second method) uses exactly the same default assumptions as conventional methods, including data selection, but expedites the process by relying on an existing data base and by not spending time documenting procedures not followed.

In addition to scientific soundness, there are a number of policy reasons for adopting an expedited procedure. Either suggested procedure will greatly expedite potency assessments. The CEPA derived about 200 potency estimates (125 new ones plus 77 which were then compared with existing potency estimates by the CEPA) from October 1990 to April 1991 (and they were not committed to working full-time on this project only). This compares quite favorably with up to five person-years per potency estimate under conventional risk assessment procedures. And the CEPA could easily have done more if there had been information on other substances in the data base.

Expedited risk assessments will better protect the public. If we know the public is exposed to carcinogens but do not have a sense of the magnitude of the risk, we do not know how great a problem the substances pose. Furthermore, if there are genuine harms from unassessed carcinogens, but we are unaware of them, the public bears the burden and the costs of our ignorance. It is one thing to withhold judgment on scientific matters when doing research, it is quite another to withhold judgment when harmful public health consequences may result. Remaining ignorant about known toxic substances and failing to regulate them have substantial social costs. In clinical medicine, physicians are aware of the costs of inaction; in environmental toxicology we need a similar awareness and action to avoid these public costs.

(Even someone who may not have strong health-protective concerns should find the expedited procedures attractive. It is important not only to know of the existence of carcinogens, but also to have some sense of their potency, for then one has a somewhat better sense of whether or not a health risk exists. If one has both potency and exposure information, one can assess public health risks at least in an approximate way. Thus, even for those who may resist regulation, knowledge of potencies is a first step in an informed public health policy.)

Risk assessments, however, pose the possibilities of mistakes. And, while it is clear that failing to assess the potential harms from carcinogens imposes costs, expedited procedures may not assess the risks from substances quite as carefully as more conventional procedures. Certainly this is a possibility. To account for the costs associated with the possibility of errors, a simulation analogous to Figure 4–1 compares the costs of mistakes of two expedited procedures considered for regulatory action in California (and just discussed) with conventional risk assessment procedures.

Figure 4–4 shows that the total social costs of conventional risk assessment procedures are in general much greater than the costs of the TD_{50} most sensitive species or the LMS default method.[172] Even the least "accurate" expedited procedure, the TD_{50}, which results in major overregulation or underregulation 25% of the time, is preferable to conventional quantitative risk assessment procedures as long as the health costs of underregulation are more than 250% of the economic costs of overregulation. The cost advantage for the LMS procedure continues until the costs of overregulation are 500% of the costs of underregulation. Commentators in the field appear never to assume on average such unfavorable cost ratios.[173] Thus, it appears that approximation methods will always have a cost advantage for reasonable ratios of regulatory costs and benefits and reasonable error rates. The more substances that are evaluated reasonably carefully using approximation procedures, the lower are the social costs.

The simulation provides strong economic and public health arguments for adopting expedited approximation risk assessment procedures. The costs assigned to underregulation and regulatory false negatives are an attempt symbolically and quantitatively to capture in an approximate way the public health costs of disease and

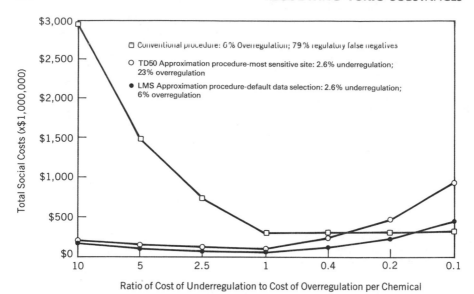

Figure 4–4   Comparison of Total Social Costs of Evaluating 369 Proposition 65 Carcinogens: Conventional Potency Assessments vs. Two Expedited Procedures Used by the California Environmental Protection Agency. [From C. Cranor, "The Social Benefits of Expedited Risk Assessments" (forthcoming).]

premature death that might be caused by unregulated or underregulated carcinogens. Conventional methods are very expensive when all social costs are taken into account. The costs from any inaccuracies in the expedited procedures are greatly outweighed by the gain in assessing a larger universe of substances.

Moreover, the cost figures would have been somewhat higher had we incorporated the costs of agency time and resources devoted to conventional procedures compared with those of expedited methods, for the former are much more labor and resource intensive.[174] Thus, expedited procedures have an advantage because they do not have the opportunity costs of conventional procedures (i.e., substances left unregulated) and the costs for evaluating each substance would be much lower.

(Other commentators have demonstrated that there are substantial costs associated with the regulatory system's reliance on time-consuming animal bioassays to *identify* potential human carcinogens, because many carcinogens escape regulatory attention. The scientific challenge is to develop approximation methods to expedite the *identifications* of carcinogens so we do not have the social costs of unidentified toxins. The regulatory challenge, then, is to adopt such methods for regulatory purposes. I have not argued for these points, but others have.)[175]

We pay a high price at present for scientific care in conventional risk assessments. Potential medical and death costs are incurred because we do not have a sense of the potencies of (and probable exposures to) identified carcinogens. The costs of unregulated substances exist but stay hidden until an epidemiological study identifies a carcinogen as a cause of a cancer cluster or until an agency otherwise gets around to evaluating and regulating the substance. Expedited procedures would enable us to assess the potency of carcinogens, to have a better evaluation of health risks they pose, and to take regulatory action if necessary.

The preceding discussion compares the total social costs of two expedited potency assessments with a conventional procedure when there are no complicating real world features. A second comparison to accommodate a possible concern about any potency assessment based on animal studies would take into account the claim made by some that animal studies are no more than 70% accurate in predicting carcinogens for human beings. Consequently, a second set of equations and the corresponding graph would take into account both of these complicating factors. The result is illustrated in Figure 4–5.

The total social costs from conventional procedures drop because only 70% of the universe of 369 substances are assumed to be true carcinogens and thus pose a real risk to human beings. There is a similar discounting procedure that applies to the expedited methods although the results are not quite as dramatic.

The shapes of the curves change somewhat with the expedited procedures showing a faster rise in total social costs when the individual costs of overregulation are assumed to be higher than the individual costs of underregulation. The reason is that 30% of the identified carcinogens are assumed to be false positives. In addition, the crossover points shift to the left, but as long as the costs of underregulation are greater than the costs of overregulation, expedited procedures are superior to science intensive conventional approaches.

Nonetheless, there is an important lesson in this figure. The current presumption

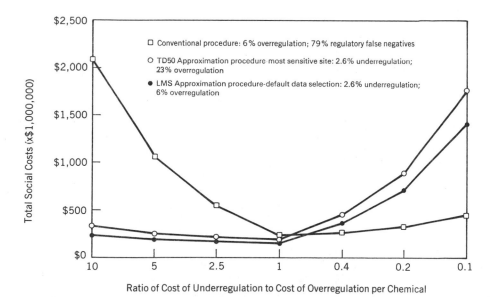

Figure 4–5 Comparison of Total Social Costs of Evaluating 369 Proposition 65 Carcinogens: Conventional Potency Assessments vs. Two Expedited Procedures (Assumes 30% False Positives for Identified Carcinogens). [From C. Cranor, "The Social Benefits of Expedited Risk Assessments" (forthcoming).]

in agencies appears to be that science-intensive care-by-case assessments are needed for every substance. However, the preceding results (Figures 4–4 and 4–5) suggest, to the contrary, that such assessments are necessary only if there is low human exposure and if the costs of regulating a substance are high compared to the risks to human health. Regulatory procedures, thus, should be refined to make discriminations between substances in order to utilize agency and social resources more efficiently and to better protect public health.

Expedited risk assessments can be used for a variety of regulatory purposes: to post warnings (as under California's Proposition 65 or OSHA's right to know provisions), to require the best available technology to reduce the risks of exposure (as under the Clean Water and Clean Air acts), or even to set ambient exposure levels (as under the OSH Act). The last is typically the most difficult, for it requires a specific exposure level be set as a matter of law and typically receives close adversarial scrutiny during the regulatory process and often in the courts. Even in this case, however, expedited procedures can be utilized as long as companies have an adequate appeal process for amending the regulations as new or better information becomes available. I have not addressed these regulatory issues because they would require extensive treatment that is beyond the scope of this work.

The upshot of the preceding three sections has been the conclusion that agencies should take into account the social consequences of adopting one risk assessment design and procedure rather than another to regulate carcinogens. The social consequences should include concern with promoting—rather than frustrating—legal goals and most especially should include concern with the rate of risk assessments. The consequences of expediting risk assessment procedures appear to be on balance far better than the present science-intensive procedures that agencies use. Furthermore, expedited risk assessments are likely to have an even greater advantage when contrasted with some of the *recommended* science-intensive approaches, which aim to reduce the costs of overregulation and false positives further while promising to be even slower than present practices.

Finally, many of the recommendations made in this section address scientific uncertainties and lack of knowledge—imperfect scientific understanding. At present scientists do not know enough to identify all toxic substances accurately and to assess their potency rapidly and accurately. In the future scientific knowledge may be much better: there may be many fewer uncertainties and we may have much better understanding of the biological mechanisms involved. At that time some of the above recommendations will be moot.

Nonetheless, even if we had perfect scientific understanding and knowledge, the points made above about the *rate* of identification, assessment, and regulation remain. Even perfect science could be time-consuming and labor-intensive when applied to toxic substances. If people will be exposed to toxins, then, through their morbidity and mortality they will bear the costs of inaction and sluggish identification, assessment, and regulatory processes. Thus, even when we understand biochemical and biological mechanisms perfectly, if applying this knowledge in regulatory settings is slow, as it is at present, this situation will have to be remedied. The scientific and regulatory challenge will be to continue to utilize rapid and reasonably accurate approximation methods to identify and assess toxins, so the costs of scientific investigation are not unnecessarily imposed on the public.

MAKING PUBLIC POLICY ON EXPEDITED RISK ASSESSMENTS THROUGH THE AGENCIES AND COURTS

In the United States and most advanced countries environmental policy and public health protections are a function of at least three major interacting institutional forces: (1) the legislation under which an administrative agency acts (e.g., the Clean Water Act, the Clean Air Act, the Safe Drinking Water Act;) (2) the interpretation and implementation of that legislation by administrative agencies; and (3) the interpretation by the appropriate reviewing courts of the organic authorizing legislation and of the agencies' implementation of it. In a complex legal system all three institutions influence the extent of environmental health protections. Consequently, recommending a course of action for using scientific evidence in agencies also involves making recommendations about how agencies should interpret such evidence in light of the controlling legislation, how agencies should use this information, and how courts should interpret agency use of the information.

Although a number of factors in the past—legal decisions and sociological forces—may have substantially influenced agencies to use conventional detailed case-by-case risk assessments, it is not clear that current courts should or would require them. Both theoretical understandings of agency discretion and recent court cases suggest that agencies may have considerable discretion to implement congressional legislation and issue regulations within their areas of expertise. In particular, agencies appear to have sufficient discretion to implement policy-based expedited risk assessments.

The issue is framed by the standards of judicial review of agency actions. That is, when an agency issues regulations under the authority of legislative mandate, those regulations have the force of law and continue to have legal authority unless a court finds that an agency has exceeded its authority or violated federal procedures in its actions. Reviewing courts have a number of standards by which to judge whether an agency has exceeded its discretion,[176] but two are important in most cases: the "substantial evidence" test and the "arbitrary and capricious" test.

For each test the court must review the record as a whole or the relevant parts of it. The court "shall—. . . 2) hold unlawful and set aside agency action, findings, and conclusions found to be—A) arbitrary, capricious, an abuse of discretion, or otherwise not in accordance with law; . . . E) unsupported by substantial evidence in a case subject to sections 556 and 557 [of 5 U.S. Code] or otherwise reviewed on the record of an agency hearing provided by statute."[177]

Both standards of judicial review share a common feature: "The court must decide that a reasonable official could have made a particular governmental decision, given the factual 'record' in the case and the particular legal norms to be applied to it."[178] But there are several differences between them. The arbitrary and capricious test requires judges to be more deferential toward agency actions than the substantial evidence test.[179] More in the way of facts must be offered under the substantial evidence test "to provide adequate evidence to support" an administrator's decision.[180] And there may be a "difference in presumptions of validity."[181] It has been suggested that the phrase "substantial evidence" focuses on what quantum of evidence *the government* must show in order to justify a challenged agency decision. The syntax of "arbitrary and capricious" focuses upon *the citizen challenger's* burden in overturning an agency decision. In theory, governmental decisions in both cases are favored with a presumption of validity, but courts are far more willing to undertake vigorous scrutiny of the former. That willingness accounts for the difference in deference and in burden of proof echoed in the semantic labels of the tests.[182]

In reviewing the factual record an administrator has not sufficiently satisfied the conditions of the substantial evidence test by considering only evidence that favors the desired decision. The administrator must also take into account "contradictory evidence or evidence from which conflicting inferences could be drawn whatever "in the record fairly detracts from its weight.""[183]

Under the arbitrary and capricious standard the "scope of review. . . is narrow and a court is not to substitute its judgment for that of the agency."[184]

> Normally, an agency rule would be arbitrary and capricious if the agency has relied on factors which Congress has not intended it to consider, entirely failed to consider an important aspect of the problem, offered an explanation for its decision that runs counter to the evidence before the agency, or is so implausible that it could not be ascribed to a difference in view or the product of agency expertise.[185]

The substantial evidence test is required by the Occupational Safety and Health Act for court review of OSHA actions. The Supreme Court's *Benzene* decision interpreting an OSHA action under that statute increased agency use of risk assessment. Subsequent to that case lower courts differed over the degree of judicial deference warranted by the substantial evidence test in reviewing OSHA decisions.[186] Some interpreted it as requiring a closer review of OSHA's decisions; some a more deferential review. Early on it was not clear whether the more demanding or the more deferential review would predominate in the federal courts.

At the time of the decision it appeared the Court would follow the strict interpretation. It now appears that both the Supreme Court and the circuit courts of appeal are giving OSHA greater rather than less discretion in setting workplace health standards. This trend should be strongly encouraged, not only for OSHA, but for the other environmental health agencies as well.[187]

Commentators differ in their judgments about the extent of judicial deference to administrative actions. One author citing several others notes that the D.C. Circuit at least up to 1988 had been increasingly active and less deferential in reviewing agency actions.[188] This has increased the time required to make policy through rulemaking as well as the costs to the agency.[189] Arguing that this is undesirable, however, he urges court deference. He also cites a number of authors who indicate that on theoretical grounds "agencies enjoy significant comparative advantages over other institutions of government as sources of policy decisions."[190] And he recommends "greater deference to majoritarian institutions of government [which includes administrative agencies whose top personnel are subject to appointment by elected officials] and adherence to judicial precedents"; both, he claims, for the most part "converge toward recommending judicial deference to agencies."[191] A second commentator does not applaud increased judicial deference to agency actions.[192] He thinks that the deference results from an interest in promoting marketplace solutions to regulatory problems and argues that this is not always appropriate for environmental health regulations.[193]

A review of some recent circuit court decisions indicates that courts appear to be more deferential to agency decisions.[194] Dwyer, noting several recent decisions, indicates that lower courts "have been quite deferential to OSHA's judgments regarding which risks are 'significant,'"[195] and several recent Supreme Court and circuit court decisions support this view.

In 1982 the Supreme Court in *Baltimore Gas and Electric Co. v. NRDC* held that the Nuclear Regulatory Commission did not act "arbitrarily and capriciously" in making the assumption which involves substantial uncertainties that the storage of specified nuclear wastes "would have no significant environmental impact . . . and thus should not affect the decision whether to license a particular nuclear powerplant."[196] And it added that reviewing courts

> must remember that the Commission is making predictions, within its area of special expertise, at the frontiers of science. When examining this kind of scientific determination, as opposed to simple findings of fact, a reviewing court must generally be at its most deferential.[197]

In the next term the Supreme Court upheld the EPA's discretion in defining "stationary source" for purposes of implementing the Clean Air Act.[198] The Carter administration had placed one interpretation on this term which Congress had not clearly defined. When the Reagan administration took office it changed the definition of the term and modified the EPA's implementation of the Clean Air Act. The Court held that

> an agency to which Congress has delegated policy making responsibilities may, within the limits of that delegation, properly rely upon the incumbent administration's views of wise policy to inform its judgments. While agencies are not directly accountable to the people, the Chief Executive is, and it is entirely appropriate for this political branch of the Government to make such policy choices.[199]

One commentator has suggested that if agencies are given discretion to interpret congressional legislation, they have equal or greater discretion to issue regulations.[200] Commentators appear to agree on the importance of this decision for judicial deference to agency actions, whether they find the decision desirable or undesirable. Agencies thus appear to have considerable discretion to interpret their legislative mandates and issue regulations under them.

Circuit courts of appeal, following the Supreme Court's lead, have also shown substantial deference to agencies' regulations. In 1984 the Circuit Court of Appeals for the Ninth District, reviewing OSHA's regulations for maximum permissible exposure levels of airborne arsenic against industry challenge, upheld the agency.[201] The Chemical Manufacturers of America (CMA) challenged OSHA's significant risk determinations as well as its regulation of a particular form of arsenic. ASARCO challenged OSHA on its feasibility findings. CMA argued that alternative inferences should be drawn from the evidence before OSHA and claimed that OSHA had failed to justify its position on substantial evidence in the record. The court disagreed, saying that it "must defer to OSHA" in such circumstances.

Of particular interest is CMA's challenge to OSHA's risk assessments. The court upheld OSHA's reliance on "a linear, no-threshold, cumulative dosage model that CMA argue[d] incorrectly determines that there is no safe level of exposure to arsenic."[202] It further found that the agency had properly rejected various alternative scientific studies on appropriate grounds. Thus, the court deferred to OSHA's expertise and policy judgments in performing its risk assessment. The court suggests that such decisions are clearly within the agency's discretion as long as reasons are given for decisions made.

In 1986 the Circuit Court of Appeals for the District of Columbia, reviewing OSHA's long-delayed regulation of ethylene oxide (EtO), supported OSHA's long-term exposure limit against industry challenge but not its refusal to issue a short-term exposure limit against challenge by Public Citizen Health Research Group.[203] An industry group, the Association of Ethylene Oxide Users (AEOU), claiming that no individual piece of evidence proves a relationship "between EtO exposure and various health effects," challenged many features of OSHA's decision: OSHA's use of rat studies (in addition to one human epidemiological study), its choice of mathematical models to extrapolate risks to human beings at levels where no data exist, and its assumption that EtO is harmful at the lowest concentrations and does not operate via a threshold mechanism.[204] The court rejected this challenge and noted that the agency must be given "leeway when regulatory subject matter is not subject to strict proof one way or the other . . . [for otherwise] *strict proof would fatally cripple all of OSHA's regulatory efforts* and run counter to the legislative branch's express delegation of hybrid rulemaking power to OSHA."[205]

Finally, in 1988 the Court of Appeals for the District of Columbia, reviewing OSHA's revised standards governing workers' exposure to asbestos, upheld OSHA's regulations against industry challenge on two of three major issues.[206] The court upheld OSHA's technical and policy decisions concerning exposure duration, the inclusion of smokers in epidemiological studies used to set the revised standards, and the gap between compliance levels and OSHA's permitted exposure level (PEL) used in determining its "significant risk" level.[207] Most important it upheld OSHA's technical risk assessment assumptions including use of the linearized high-dose to low-dose extrapolation model and use of "conservative assumptions in interpreting the data with respect to carcinogens, risking error on the side of over-protection."[208] While it expressed some worry that this sometimes might have the perverse effect of "*increasing* rather than decreasing risk," it deferred to OSHA's decision.[209]

These cases taken together indicate considerable court deference to agency actions based on fairly standard risk assessment procedures. They also suggest that the courts would defer to agencies' expertise and policy judgments if they chose to fulfill their legislative mandate by using somewhat different risk assessment methods which expedited the evaluation of substances as long as the agency had good reasons for this approach. This issue has not been litigated, however. Nonetheless, court language indicating deference both to agency expertise on the frontiers of scientific knowledge and to an administration's "wise policy" suggests that agencies might have substantial latitude to fulfill their legislative mandate with policy-guided, public health sensitive, and expedited risk assessment procedures. Agencies should argue that even though expedited risk assessments might be somewhat less science intensive than the conventional, time-consuming procedures typically followed, they are scientifically sound and their results are very close to conventional procedures. The time spent on conventional risk assessments results in substantial underregulation, a result the court should wish to avoid. Expedited procedures greatly reduce underregulation, thus enabling agencies better to fulfill their health-protective legislative mandates and reduce the total social costs from human exposure to carcinogens.

CONCLUSION

In this chapter as in Chapter 2, I have argued that we face a paradigm choice in how scientific evidentiary procedures are treated in the law. Present agency practices and several recommended alternatives to them aspire to the ideals of research science in assessing the risks posed by each substance. This is a mistake. Attempts to reduce uncertainty by aiming for greater accuracy for each substance to ensure that the demanding standards of science are met as much as possible and to perform exceedingly careful risk assessments will all exacerbate current problems.

The identification, assessment, and regulation of potential carcinogens are all too slow to evaluate adequately the existing universe of 50,000–100,000 chemical substances, and the 1000–1500 new ones that are added each year. Thus, for identified carcinogens we need to adopt something like the expedited approximation procedures I have indicated, which permit faster evaluation of potentially toxic substances. (We also need faster procedures to identify toxic substances, but I have not argued in detail for this.) The way to do this is to shift our paradigm of the needed evidence and the amount of attention and documentation each substance deserves. What is required is not a closer approximation to ordinary research science but quicker, more sensitive health-protective approximation procedures that are accurate enough for the regulatory purposes in question. The challenge to the scientific community is to develop or refine such procedures for risk assessment purposes. The philosophic challenge to the scientific and legal community is to see the importance of doing this and to endorse a different paradigm.

Agencies appear to have sufficient flexibility under current laws to develop such procedures because reviewing courts in recent decisions seem to be giving greater deference to the agencies. This is salutary. If we shift our idea of the appropriate scientific procedures so that the relevant institutional and moral norms shape the kind and degree of scientific accuracy we need, we will be better off. The law and the relevant moral concerns should determine the epistemic procedures we use, not the other way around.

5

Epistemic and Moral Justification

My husband worked in the cotton mill since 1937 to 1973. His breath was so short he couldn't walk from the parking lot to the gate the last two weeks he worked. . . .

He was a big man, liked fishing, hunting, swimming, playing ball, and loved to camp. We liked to go to the mountains and watch the bears. He got so he could not breathe and walk any distance, so we had to stop going anywhere. So we sold our camper, boat and his truck as his doctor, hospital and medicine bills were so high. We don't get to go anywhere now.

The doctor said his lungs were as bad as they could get to still be alive. At first he used tank oxygen about two or three times a week, then it got so he used more and more. So now he has an oxygen concentrator, he has to stay on it 24 hours a day. When he goes to the doctor or hospital he has a little portable tank.

He is bedridden now. It's a shame the mill company doesn't want to pay compensation for brown lung. If they would just come and see him as he is now, and only 61 years old. . . .

Mrs. Steve Talbert
Charlotte (N.C.) Observer, February 10, 1980

I have argued that we should avoid the temptation to adopt the ideals of research science in torts and administrative law because this tends to lead to an excess of false negatives and to underregulation of carcinogens as well as to the undercompensation of plaintiffs in tort cases. Instead, scientific practices and evidentiary standards should be sensitively tailored to serve the goals and aims of these legal institutions or, more simply, legal norms should guide epistemology. Adoption of different standards in the law might in some cases lead to mistakes of overregulation or overcompensation, but the overall social costs would be lower, because a better balance of false negatives and false postives would be found. Ideally, risk assessment and regulatory procedures would impose no costs on anyone, however because of imperfections in knowledge and because of uncertainties, the choice of evidentiary standards is in effect a choice between imposing overregulation, overcompensation, and their associated costs, and imposing health and other costs on individuals because of underregulation and undercompensation. Beyond the law, however, I will argue that justice (and distributive considerations more generally) requires that priority be given to avoiding the latter.

This chapter focuses on two parts of this argument that need to be developed further: (1) the standards of evidence ought to be appropriate to the institutional context and (2) justice requires that priority be given to avoiding false negatives and underregulation. One requires justification of the epistemic presupposition, the other, justification of the underlying moral view.

EPISTEMIC JUSTIFICATION

The argument presupposes in critical places that different standards of evidence (or different burdens of proof) are appropriate for different areas of human endeavor. This may appear obvious, and I have tried to illustrate its defensibility in detail with respect to the tort and regulatory law. Nonetheless, some discussion of the general claim is warranted given its importance to the overall argument. What is the connection between burdens of proof and institutional norms?

The burden of proof that must be satisfied in order to make a particular institutional decision will depend at least in large part upon the kinds of mistakes one seeks to avoid. Some institutions will aim to provide greater protections against certain kinds of mistakes; some will protect against other kinds of mistakes. The criminal law stringently protects against convicting innocent people. To avoid this, the equivalent of false positives, the state must overcome a high burden of proof to establish its case. By contrast, in screening patients for life-threatening diseases we might seek very much to avoid missing someone who has the disease, to avoid false negatives. Thus, we would want to make it very easy to detect the presence of disease, even if this meant in some cases wrongly identifying some as diseased who were not. This approach would be especially attractive if we had a benign treatment for all identified as diseased.

Moreover, mistakes impose costs upon someone. Thus, in the design (or in the evaluation) of institutions one must consider the larger consequences of evidentiary procedures. If there will be mistakes, on whom should these costs fall? How should they be distributed?

Both the distribution of costs and the kinds of institutional mistakes one seeks to avoid, however, in turn depend upon what is at stake—the values the institution or activity seeks to protect. Institutions embody a variety of values that help set the standard of proof needed to reach a decision. In some cases the values tend to be few and obvious; in others they express a complex matrix of concerns, some of which may be less easy to discern. In research, scientists seek primarily to add to the stock of human knowledge, and they have devised research strategies, inferential standards, and the peer review process to prevent accepting false information as true. This is not the only value, but a very important one. The criminal law seeks to avoid wrongly punishing innocent people because of the particular injustice this inflicts upon them. It also seeks to equalize to some extent the competition between the state and individuals. For these reasons, among others, criminal procedure and evidentiary rules impose high burdens of proof on the state in order to secure these values and prevent false positive mistakes. The Food and Drug Administration has quite different evidentiary procedures for approving drugs and food additives as safe and efficacious, because it aims to prevent unsafe or ineffective food and drugs from entering commerce. It seeks to protect against mistakenly finding a substance safe when it is not; it protects against false negatives. Recently, the FDA expedited its approval process for AIDS pharmaceuticals in order to make available drugs that might be effective against this syndrome. The rationale appears to be that it is better to risk marketing a drug that might be effective against a fatal disease than to delay approval until the agency is certain of the drug's safety and effectiveness. In short, the urgency of trying to save lives justifies taking some chances that the drug may not be wholly effective or totally safe. But even if it is not totally safe, if it effects a cure or prolongs useful life, it can hardly be worse than the disease.

In another context, screening women for rubella antibodies in order to prevent congenital rubella, doctors incur a modest number of false positives in order to be sure to prevent the devastating effects of the disease. "The inefficiency of vaccinating a number of women who are already immune is a small price to pay."[1] In screening for the HIV antibody doctors stringently protect against false negatives, thus making the test more sensitive but incurring more false positives. Positive screening results must be further tested to eliminate false positives.[2]

These examples and the preceding argument suggest that the presupposition of much of the book, which seems obvious, is straightforward: the norms and goals of the institution guide the kinds of mistakes the institution seeks to avoid, which in turn determine the burdens of proof that are adopted for decision making within the institution. Burdens of proof determine how easy or how difficult it will be to make certain kinds of mistakes in institutional decisions.

In addition to this general presupposition, the argument of the book has relied upon several more specific claims. One is that there should be some "appropriate" balance struck between false positives and false negatives that result from tort law and administrative agency decisions. Again this seems clear, but several considerations support this point. For one thing, a decision to postpone action is just as much a decision under uncertainty as a decision to take precautionary action in the meantime. This has long been recognized in clinical medicine. In assessing potentially toxic substances a decision not to act risks false negatives, whereas a decision for precautionary regulation risks false positives. Since one risks both mistakes, both should be acknowledged and the decision process designed to find an appropriate balance between them.

Neither of two extreme options has been urged. On the one hand, I have not argued that regulatory agencies should always make the precautionary decision to regulate or that tort courts should always award compensation when an injury from a toxic substance is suspected. These would always prevent false negatives, but at the cost of false positives. Both strategies would be analogous to Pascal's Wager—that because of the importance of what is at stake one decision alternative always dominates the other. On the other hand, I have not argued that agencies should never act or that tort courts should never award compensation. This would always prevent false positives at the cost of false negatives. Quite the contrary: it seems agencies and tort courts are tending to emphasize such concerns too much and should have greater concern with false negatives. Thus, I have not suggested simple decision rules but modifications in complex existing or recommended procedures in institutions to try to achieve a more appropriate balance of these kinds of mistakes. There are a number of considerations that argue for finding some balance between the kinds of mistakes.

If one is concerned to minimize the social costs of errors, a widespread assumption, then one must take into account both kinds of mistakes. The appropriate balance, then, is one that minimizes the costs of the two taken together. Thus one should minimize the sum of the cross-product of the cost of each mistake times the number of such mistakes. Expressed symbolically this is

$$\min \left[(C_{FP} \times N_{FP}) + (C_{FN} \times N_{FN}) \right] \tag{1}$$

where C = cost, N = number, and FP is false positive and FN is false negative, as usual. If the costs of the testing procedures are not zero, then these should be included as well.[3] Symbolically we can represent the total cost of a decision procedure as

$$\left[(C_{FP} \times N_{FP}) + (C_{FN} \times N_{FN}) + C_T \right] \tag{2}$$

where C_T = cost of the requisite institutional tests including the costs of human resources.

If we consider the universe U of chemical substances that might be carcinogens, and p_c as the probability that a substance is a carcinogen (and $1 - p_c$ = probability that it is not a carcinogen), then the number of false positives is $(1 - p_c) \times \alpha \times U$, and the number of false negatives is $p_c \times \beta \times U$, where α is the chance of a false positive and β the chance of a false negative which results from the testing procedure. Thus, 2) becomes

$$\left[(C_{FP} \times (1 - p_c) \times \alpha \times U) + (C_{FN} \times p_c \times \beta \times U) + C_T \right] \tag{3}$$

If we seek to minimize the total cost, then we take the minimum of this sum:

$$\min \left[(C_{FP} \times (1 - p_c) \times \alpha \times U) + (C_{FN} \times p_c \times \beta \times U) + (C_T \times U) \right] \tag{3a}$$

Since the total number of substances to be "tested" is common to each term of the sum we can ignore it, and we get

$$\min \left[(C_{FP} \times (1 - p_c) \times \alpha) + (C_{FN} \times p_c \times \beta) + C_T \right] \tag{3b}$$

which expresses the way to minimize the cost of testing each substance.

Thus, the costs are merely a function of the costs of false negatives and false positives, the probability that a substance is a carcinogen and the error rates of the procedures for discovering whether a substance is a carcinogen or not for regulatory or tort purposes.

The strategy for a particular institution, then, is to have "test" procedures, that is, evaluation procedures, which, given the respective costs of false negatives, false positives, and the procedures themselves, minimize the total costs of evaluating the universe of substances.

In the argument of the book I have largely ignored the costs of testing, except for the agencies' opportunity costs from not assessing known carcinogens. To oversimplify a bit, throughout the book I have noted how our evidentiary procedures can increase or decrease the kinds of errors we make—the α and β errors in evaluating substances for legal purposes.

For the tort law I showed how subscribing to the demanding evidential standards of typical research science in conjunction with the plaintiff's burden of proof could greatly increase the false negatives as long as some cases have merit [because in accordance with (3b), β is greatly increased] The reason for this is that even a meritorious tort case could not "get going," because the plaintiff could not satisfy the burden of proof.

For the regulatory law I argued that because we have a universe of known carcinogens, demanding standards of evidence and perceived legal requirements imposed on carcinogenic potency evaluations produced high opportunity costs from the evaluation because so few carcinogens were assessed (in effect, β was quite high for the universe of known carcinogens as a whole). Faster but approximate methods are much more defensible because some overregulation, a relatively low-cost mistake, is incurred to reduce underregulation, a higher cost mistake. Conventional assessment procedures would be even more costly if the human and dollar costs to the government of evaluating substances were also taken into account. Finally, using similar arguments, others have shown that some scientific tests or procedures used to identify carcinogens, may be so expensive as to preclude their use totally.[4]

The symbolic characterization of the appropriate balance of false positives and false negatives thus enables us to understand fairly precisely where the problems lie in the different areas of the law regulating toxic substances.

The particular goal of minimizing the costs of mistakes is one with which economists are especially concerned and one which presupposes a particular approach to collective decision making, since economists seek to maximize social benefits, and mistakes frustrate this aim. But this aim need not be restricted to them. It is one important consideration by which to judge any institution, since, except for any distributive effects, we are all better off if such costs are minimized. However, even if one does not explicitly seek to minimize the costs of mistakes but is merely concerned with more than one kind of mistake, then both false positives and false negatives should be taken into account in designing institutional procedures. If one did not seek to minimize the cost of errors but sought some other outcome, one would need a different analysis.

The "appropriate" balance, given the arguments of this section and throughout the book, depends upon the institution in question. And, of course, one would not necessarily strike the same balance between false positives and false negatives from one institution to another. Scientific practices balance FPs and FNs in one way, while different balances should be struck in the regulatory and tort law. Both place a much greater concern on avoiding FNs than exists in scientific practices.

In Chapter 2 I argued that in the tort law the costs of false negatives (C_{FN}) and false positives (C_{FN}) are approximately equal, but that demanding scientific standards of evidence distort this relationship, since in science $C_{FP} \gg C_{FN}$. However, by conceding that $C_{FN} = C_{FP}$ for torts I may concede too much, since at least some tort law principles suggest that $C_{FN} > C_{FP}$.[5] In Chapter 4 I indicated that for regulatory purposes the costs of FNs are typically greater than FPs, even up to ten times (or more) larger. This appears to be a fairly common assumption. Of course, the C_{FN}/C_{FP} ratio may differ between regulatory institutions, since different agencies operate under different statutes.

Not only are scientific burdens of proof at odds with evidentiary standards we should use in administrative agencies or the tort law, but the mistakes scientific and legal institutions seek to avoid are incompatible. And if scientific practices dominate in legal institutions, we are inadvertently making a particular moral/value choice. We not only give priority to avoiding one kind of mistake, we are permitting the concerns of research science—factual accuracy to avoid false positives—to take precedence over other matters of importance—protecting public health. This is a social choice and the choice could be made differently. Identifying the relation between evidentiary standards and the norms of the institutions in which they are used makes it clear that the choice of evidentiary standards is a normative choice. What norms should we foster? Which are important in designing the institutions in question? Typically, such choices may not have been consciously made but are hidden behind standard practices. One aim of this book has been to expose these implicit practices and to argue that the choices should be consciously made, as if we were consciously designing the institution.

The presumption about the importance of avoiding false negatives and under-regulation has sometimes been defended in terms of the goals of the tort law or environmental health law, but they have a deeper justification, I believe; they are a matter of justice.

MORAL JUSTIFICATION

The relative importance of avoiding different kinds of mistakes is difficult to specify with precision. In the law such issues rarely receive systematic treatment. Even though legislation for the administrative agencies may express to some extent the community's ideals,[6] it is typically not developed as a result of a consistent theoretical view but is the product of conflicting political forces. In the common law of torts distributive issues are not decided as a matter of consistent theory but are developed by means of case-by-case adjudication between two parties. Of course, judges try to produce consistent decisions over time, but they do not always succeed.

If the law is not necessarily consistent, it is especially important to have a better view of more fundamental and consistent normative principles to guide our epistemology. And this need is even greater because we must try to control exposures to toxic substances under considerable uncertainty. In regulating toxics "through a glass darkly," we need to be clearer about the normative considerations that can aid us in this endeavor.

For one thing, social decisions are the product of both the relevant facts and normative considerations. Even when facts are known with certainty we must rely upon norms to guide our decisions. When the facts are even more uncertain than usual, as they are for toxic substances, norms become more important, for we have to place greater reliance on them. If we have to make a choice under great uncertainty our only or main guidance may be the concerns we promote or frustrate as a result of the decision. The uncertainties focus our attention more clearly on the norms at stake when we decide to regulate or not. (And even a decision to do nothing in the face of uncertainty is an important normative decision.)[7] Thus, if epistemology under uncertainty is to receive greater guidance from normative considerations, we should be aware of what the normative principles commit us to. Moreover, different normative principles will assign different weights or urgency to avoiding false negatives and false positives.[8]

In addition, as Annette Baier noted, moralities select "which harms to notice and worry about where the worry takes the form of bad conscience or resentment."[9] If this and the preceding claim are correct, then we should understand the implications of different moral views, so we see both what they regard as particularly urgent and what harms we should worry about. Once we are clearer about this we may be able to make a better choice between principles.[10]

Moreover, for these purposes we need principles to distribute the costs of mistakes; we need principles of justice. Not only are such principles concerned with distribution, but if justice is the first virtue of institutions, as I believe it is,[11] we should want our environmental health institutions to be just.

And, finally, since normative considerations are important for decisions under uncertainty, it is at least possible that there might be greater agreement on them than on the uncertain facts. This may well be too much to hope for, but agreement on distributive principles could, I believe, greatly reduce many of the disagreements concerning risk assessment and risk management, since those disagreements are frequently driven by normative considerations.[12] However, whether or not greater agreement on norms is likely, we need to understand and evaluate some of the leading normative principles in order to try to achieve as much agreement as possible on the just treatment of the people affected by our decisions.

The following discussion considers four principles for addressing some of the implicit distributive questions that arise in torts and in administrative law. These accounts place different degrees of urgency on environmental health harms and identify different harms to notice and worry about. I begin with the distribution implicit in the Occupational Safety and Health Act. Next, I consider two theoretical moral views for such distributive concerns. Finally, I discuss the attractive features of "distributively sensitive" theories for addressing some of the distributive questions posed by the normative concerns mentioned.

In considering these issues, I focus for the most part on workplace health protections as a relatively manageable example of environmental health protections.[13] To what extent should a society spend its resources to protect workplace health as opposed to spending resources to create additional jobs, to preserve existing jobs, to increase general societal wealth, or to preserve existing industries and business? How urgent are these different interests? To what extent are they worthy of noticing and worrying about?

These are versions of the questions I have posed throughout and they are some of the major concerns arising out of recent workplace health litigation related to the Occupational Safety and Health Act of 1970.[14] Discussion of this act and litigation concerning it shows one attempt by Congress and the courts to adjudicate some of the issues suggested above. Implicit in the statute and the legal cases is a theory about the distribution of resources to these different activities. This slice of federal law does not provide adequate treatment of the issues involved; thus we must go beyond it to consider deeper theories of justice in deciding such tradeoffs. In turn, we must consider the adequacy of these theories for deciding such matters.

The creation of jobs is important, for job opportunities provide one of the main means by which citizens can secure for their own welfare. This is especially true in an individualistic and capitalistic economy, in which individuals may have to fend for themselves and in which there may be no central planning agency to provide both job opportunities and a welfare floor. However, the need to provide jobs may obscure their nature. Employees, in providing for themselves, often have faced coerced choices and have had to trade away their good health in order to have an income above the welfare minimum, if there has been one. For example, OSHA has found (1) that the "rate of mortality from lung cancer among employees working on top of . . . coke oven batteries for 5 years or more [was] 10 times greater than normal,"[15] (2) that in a 3-year period, 13 workers in the plastics industry died of a very rare angiosarcoma of the liver due to polyvinyl chloride exposure,[16] (3) that 3 million to 5 million workers were exposed to high concentrations of asbestos, a potent carcinogen,[17] resulting in 99 excess deaths (above the rate of mortality for the U.S. white male population) out of 632 workers,[18] (4) that 1 in 12 workers exposed to cotton dust, some 35,000 total, suffered the most disabling form of byssinosis, or brown lung,[19] (5) that at 500 micrograms per cubic meter exposure to arsenic (the previous OSHA standard) there were 400 excess deaths per 1000 people,[20] and (6) that female nurses exposed to ethylene oxide (EtO) had a statistically significant increase in spontaneous abortions.[21]

In a just society, it seems persons should not have to accept considerable risks to their health simply to have an income above the welfare minimum. In the extreme, a person who has to choose between starving and working in an industry that is likely to cause premature death does not have a legitimate choice. Even if one chooses to work in such a risky job, one is not necessarily treated justly on the principle that "to he who consents no injustice is done,"[22] for there is not legitimate consent. Even if a person faces the somewhat less extreme choice between living in poverty and taking such a risky job, consent does not show he is treated justly. Such choices force people to choose between having a survival income, or worse, together with any loss of self-esteem and other psychological consequences, or to take substantial chances of contracting serious or life-threatening diseases in order to improve their lot. A just society would not force this Hobson's choice.[23]

In an individualistic society in which personal choice and autonomy are valued, it is especially important to have a just background which conditions choices about careers and life plans. Institutions and social practices importantly shape persons' lives. If these institutions force people into unjust relationships, it is not clear how much moral credence we should give their choices in such circumstances. The notions of legitimate choice and autonomy are empty if the alternatives open to people are too constrained or if they present only undesirable possibilities. Thus, although consent to working conditions historically has been an important consideration in judging the justice of working conditions, I do not discuss it here, since background considerations of justice seem of greater importance.[24]

In the discussion that follows, I consider some principles for judging background institutions of society. Specifically, I consider several different principles for adjudicating the tradeoff between protecting workplace health, creating job opportunities, and generally increasing societal income and wealth.

Principles Implicit in the Occupational Safety and Health Act

The Occupational Safety and Health Act authorizes the Secretary of Labor (the secretary) to issue "occupational safety and health standards."[25] The critical section of legislation is 6(b)(5), which directs the secretary, when issuing occupational health standards for "toxic materials or harmful physical agents," to set the standard which

> most adequately assures, *to the extent feasible,* on the basis of the best available evidence, that no employee will suffer material impairment of health or functional capacity even if such employee has regular exposure to the hazard dealt with by such standard for the period of his working life.[26]

This section contains both the substantive ideal of workplace health protection (that employees suffer *no* material impairment of health or functional capacity) and the more practical technological and economic feasibility requirements that limit pursuit of that goal. The feasibility requirements have been taken to mean that if there is not existing technology or technology that is relatively easy to develop to achieve the health protections, or if the costs are not "feasible," then pursuit of the health goal may be limited. It is largely the "economic feasibility" requirement that occasions our discussion, however.[27]

The "no material impairment" requirement does not mean that a person who is exposed to toxic substances for his entire working life must have a life expectancy identical to that of a similar person in the general population. Call this "the equal life expectancy requirement." Section 6(b)(5) recommends something short of that, for, although a person might not suffer *material impairment of health* or *functional capacity* during his working life, he might well not have a *life expectancy* equal to similar people in the general population because of his workplace exposure. Thus, the OSH Act ideal falls short of trying to ensure that we do not suffer premature disease or death because of our occupations. That act only seeks to provide for a healthy *working* lifetime. Nevertheless, the Occupational Safety and Health Act standard appears to be quite protective of health, unless such protections are not economically or technologically feasible.

The economic feasibility requirement poses the tradeoff issue of protecting workplace health and protecting the financial viability of industries and thus jobs that might be available for workers. If the costs of protecting health become too great, then it is likely that fewer people will be employed, thus producing fewer opportunities for workers to earn a living.[28] This has been addressed to some extent by the courts. The Federal Circuit Court of Appeals for the District of Columbia in the asbestos case noted that Congress recognized that "employees would not be protected if their employers were put out of business," while at the same time concluding that "standards do not become infeasible simply because they may impose substantial costs on an industry, force the development of new technology, or even force some employers out of business."[29]

The actual test for economic feasibility has yet to be fully developed by the courts. Nonetheless, the D.C. Circuit has suggested that the costs cannot be "prohibitively expensive," for Congress would not have intended OSHA to make "financial viability *generally* impossible" for a regulated industry.[30] However, it appears to be consistent with the purposes of the act to envisage the demise of an employer who has lagged behind the rest of the industry in protecting the health and safety of employees and is consequently financially unable to comply with new standards as quickly as other employers.[31] Thus the court concluded that "even if a few firms are forced to shut down, the standard is not necessarily economically infeasible."[32] In sum, if a health standard would make "financial viability generally impossible" for a regulated industry, then it is infeasible. However, merely because some (relatively small number of?) firms are forced to close because they cannot afford to comply with OSHA's mandated health standard, it does not follow that such standards are economically infeasible.[33] Furthermore, the court rejected both cost effectiveness and cost–benefit analyses as the test of economic feasibility, because they are not supported in the language of the statute or in the legislative history, and because they are especially problematic measures of economic feasibility.[34]

The implicit principles embodied in the D.C. Circuit Court's decisions suggest the following. On one hand, OSHA may set standards more stringent than existing ones in pursuit of better health for workers, unless they threaten the economic viability of an entire industry; that is too steep a price to pay for improved health. On the other hand, even the court interprets Congress as being willing to tolerate the loss of some jobs, and even some firms in an industry, if failure to impose health regulations would materially impair the health or functional capacity of workers in that industry.

These principles, although quite vague, offer some guidance for the tradeoffs between the regulatory mistakes we have considered throughout. The courts appear to place a premium on protecting against regulatory false negatives in accordance with the statute, since employees should suffer *no* material impairment of health during their working lifetime. The courts also seem to place greater importance on avoiding underregulation than on avoiding overregulation, *until the regulation threatens a whole industry*. (The evidentiary requirements considered later suggest somewhat different priorities.) Finally, the clear intention of the statute is to shift the costs of preventing workplace morbidity and mortality to the industry and the costs of its products rather than to its work force, unless the whole industry is threatened by the regulation.

However, the implicit principles might be seen as either too strong or too weak in protecting employees' health. They may seem insufficiently protective because there may be industries that pose such serious threats to health and whose products are not sufficiently attractive that the costs of implementing OSHA regulations will make the entire industry economically nonviable. We might think of this as an economically marginal but especially "dirty" industry. Nonetheless, such industries would be protected by the court's feasibility rule, and workers' health would be sacrificed to keep such industries in existence.

There is a deeper concern about the operative notion of the "viability of an industry." *Industries* are difficult to individuate. Are plastic container and metal container manufacturers part of the *same* industry or are they two different industries? If they are one, regulations of toxic ingredients in plastic but not metal containers, which might put plastics manufacturers out of business or make them less competitive, do not trigger the economic feasibility requirement. If there are *two industries* and the plastic container industry would be threatened with elimination, the economic feasibility requirement would protect it. For protecting the health of employees a broader, rather than a narrower, construal of ''industry'' seems called for under the statute and court decisions. The opposite is true for protecting industry's interests. A judicial interpretation of this critical term has not been issued.

Some might see the court's principles as too protective because they permit some job opportunities to be eliminated if a number of firms in an industry, but not an entire industry, would be put out of business. An additional concern is noted by the Supreme Court in later adjudication: the feasibility standard may be both too strong and too weak. If OSHA is permitted to issue for a single toxic substance health standards that are inordinately expensive, this may produce a severe misallocation of resources, for other substances thus may not be regulated within the feasibility requirement.[35] If OSHA has regulated some substances stringently within an industry, bringing it to the edge of financial demise, further regulation may be precluded.

The Supreme Court has not added significantly to the economic feasibility debate except to reject cost–benefit analysis definitively: it is neither required nor permitted by the Occupational Safety and Health Act.[36]

There seems to be widespread agreement among the circuit courts on the D.C. Circuit's interpretation of the feasibility requirement, although there is considerable disagreement regarding who has which burdens of proof on these matters.[37] Nevertheless, until there is further litigation at the Supreme Court level, a more definitive and specific nationwide interpretation of this phrase is left in abeyance. Present court interpretations are not fully satisfactory, for they are too vague: health protections are important and have considerable urgency, but they are greatly qualified by the economic feasibility requirement. However, neither the Supreme Court nor any circuit has definitively indicated how it would interpret the legislation to adjudicate between spending resources to create more jobs or increase wealth, but with higher risks to health, or spending resources to secure better health protections at some cost both to general wealth and to the creation of jobs. Thus there is a need for a deeper theory of justice to address this matter.

Before turning to this point we should note that to some extent the substantive ''no material impairment of health'' requirement is also greatly tempered by the evidentiary requirements of the OSH Act. That law requires OSHA to bear the burden of proving that a significant risk to health exists, and thus that regulation is warranted, and to establish its regulations by substantial evidence on the record as a whole. On traditional interpretations (considered in Chapter 4) this means that an administrator must take into account not only evidence that favors the result she thinks correct, but also ''whatever in the record fairly detracts from its weight.''[38] Thus, she must take into account evidence favoring the decision as well as evidence detracting from it and evidence favoring other decisions. If the evidence as a whole does not support a finding of a material impairment of health, then it appears OSHA may not regulate. The evidentiary requirements tend to protect against regulatory false positives while the substance of the statute protects against regulatory false negatives.

Thus, Congress granted in the substance of the OSH Act great urgency to prevent material impairment of health, which it then substantially tempered or undercut by its evidentiary requirements. On balance, then, the evidentiary requirements serve not simply the obvious health-protective goal of the act, but also other, less protective aims. As we saw in Chapter 4, the Supreme Court increased this evidentiary burden in the *Benzene* case, but more recent decisions may have relaxed it.

In sum, the obvious substantive health-protective goal of the statute is modified by the feasibility requirement, by evidentiary procedures in the statute, and by the Supreme Court interpretations. If this does not seem like a consistent approach to regulatory protections, it's not. Statutes are typically the outcome of political compromises that generate legislation in tension; this can then be modified by the courts. For a possibly better approach we turn next to more consistent approaches, philosophical principles of distribution, for two reasons: they hold out the promise of greater consistency than existing legal and political compromises and our fundamental legal institutions should be just.

Philosophical Theories of Distribution

To simplify the discussion that follows, I assume what is contrary to fact, namely, that some central planning agency has funds which can be dispensed in order to create additional job opportunities in existing industries, to institute additional health protections in those industries, to increase the general income and wealth in society (by income tax reduction or a negative income tax), or to promote any appropriate mix of these goals. Suppose also that there is an adequate welfare system to provide minimal support for those who do not currently have jobs. Suppose further there are no other uses to which this money could be put (e.g., not into education or building hospitals), so that the choice is between merely spending it on creating jobs, on improving workplace health, or on increasing the income of various groups in the community (e.g., by lowering the income tax rate or by means of a negative income tax). How such funds should be allocated is a matter to be settled by principles of distributive justice. Thus, in what follows I assume that allocations would be made from a central agency on the basis of two different distributive principles: a utilitarian principle and the "Daniels–Rawls" principle of justice. In the concluding section I consider some of the virtues of distributively sensitive theories.

Utilitarianism

The general utilitarian pattern of distribution is easy to state: one ought to allocate resources among protecting health, providing for more jobs, and increasing social income in such a way as to secure the community good or, in technical terms, to maximize utility for everyone in the community. Community utility is an additive function of the sum of individual utilities. Utility for an individual can be considered to be a mental state such as happiness or to be the satisfaction of actual desires or the satisfaction of rationally justified desires.[39] (I leave the notion of utility unspecified for most of this discussion.) A more detailed statement of the utilitarian position, however, is more complicated.

The utilitarian can look at allocation problems from two different perspectives: at the ultimate *total* utility that would be produced from a certain distribution or at the *change* in utility resulting from a certain distribution compared to a given baseline. For purposes of simplicity and in order to mirror somewhat the actual situation, I consider the problem from the latter perspective. The change of utility, compared with some preexisting baseline, is a function of (1) the change in utility resulting from increased or decreased income times the number of people affected by such changes; (2) the change of utility resulting from increased or decreased employment or employment opportunities times the number of people affected by this; and (3) the change in utility resulting from changes in morbidity and mortality times the number affected.

Symbolically we can represent total social utility generated by social policies as a function of its components generated by income, employment opportunities, and health protections distributed to people as follows:

$$\Delta U_T = f(\Delta U_I + \Delta U_E + \Delta U_H)$$

Being somewhat more precise so as to take into account the number of people affected by such changes, we should represent this equation as

$$\Delta U_T = f(\Sigma \Delta u_{Ii} + \Sigma \Delta u_{Ej} + \Sigma \Delta u_{Hk})$$

An even more precise formulation would take into account the probabilities of changes in utility for each individual:

$$\Delta U_T = f(\Sigma(p_{\Delta u_{Ii}} \times \Delta u_{Ii}) + \Sigma(p_{\Delta u_{Ej}} \times \Delta u_{Ej}) + \Sigma(p_{\Delta u_{Hk}} \times \Delta u_{Hk}))$$

where ΔU_T is the change in total utility, $\Delta U_I = \Sigma \Delta u_{Ii}$ is the change in utility for all affected due to changes in income, $\Delta U_E = \Sigma \Delta u_{Ej}$ is the change in utility for all affected due to changes in employment, and $\Delta U_H = \Sigma \Delta u_{Hk}$ is the change in utility for all affected due to changes in health (i, j, and k = individuals). And Δu_{Ii} = the change in utility for the ith person due to change in income, Δu_{Ej} = the change in utility for the jth person due to change in employment or employment opportunities, Δu_{Hk} = the change in utility for the kth person due to change in his health, and $p_{\Delta u_{Ii}}$ etc. = the probabilities of the respective changes in individual utilities. If the sum ΔU_T is greater than zero, such changes are recommended, and of the recommended changes, one ought to adopt the alternative that produces the maximum positive change in utility.

A utilitarian recommends the mix of income, employment opportunities, and workplace health protections that maximizes utility and is willing to tolerate trade-offs among all three in order to achieve this aim. Such tradeoffs are not so easily permitted by certain theories of justice, as we will see. Even though this is the utilitarian's general principle, we would like more specific guidance, and there are some suggestions. Poor health and disease are likely to produce suffering or lead to death, both disutilitarian outcomes. Good health is a nearly universal means to many ends individuals are likely to have (job opportunities, wealth, recreation, and general enjoyment). Thus, the utility of having good health compared to the utility of substantially impaired health is likely to be great.[40] Poor health also may be a great drain on personal and social resources. It is difficult to say exactly what such utilities would be because health impairment admits of degrees. It is clear, however, that very serious environmentally induced illnesses, such as leukemia (caused by benzene), angiosarcoma (caused by polyvinyl chloride), byssinosis (caused by cotton dust), and lung cancer (induced by coke oven emissions), produce a great loss of utility for the individuals involved and their families. One need only read Mrs. Talbert's letter at the beginning of this chapter to have a vivid picture of the losses occasioned by serious diseases.

Any increase in personal utility due to increased income levels also admits of degrees and depends in large measure on a person's income prior to receiving extra income: if one is poor, then a specific increment of money, say, $1000, will have a correspondingly greater impact than if one is Donald Trump.

Finally, the change in utility due to increased employment opportunities is somewhat more complex. If a person goes from being unemployed to being employed, the gain in utility can be great. If a person goes from being employed at a lower skilled job to one of higher skill and higher pay, again the utility gain can be considerable. And a person may have some increase in utility simply because there are greater employment opportunities, even though he does not immediately improve his employment. Increasing one's options, in short, can increase one's utility. There are both major insights and major shortcomings to utilitarianism.

(1) The guiding insight of utilitarianism is that it seeks to secure the maximum good for the community as a whole. This, as we saw, is a function of the good of the individuals in it. As a result utilitarianism permits tradeoffs between any social goods that will produce a net increase in social utility. This is both a strength and a weakness. Its strength is that it ensures an efficient use of resources to maximize community welfare. In principle it provides that resources will be used for whatever activity will produce the greatest community utility for a particular increment of resources.

This is also a vulnerability: it permits individual interests to be sacrificed to an efficient production of community utility. In particular, it will authorize tradeoffs between workplace health protections and increasing the general welfare, that is, increasing the income and wealth of persons in a society. Thus, for the utilitarian the choice is simply whether an expenditure of resources will promote more utility by protecting health than it will by increasing wealth—the wealth of those in the work force, the wealth of the affected industries, or the wealth of the larger society. Preventive health protections have no *special weight* in most utilitarian schemes, and utilitarians have a difficult time defending a *right* to health protections.[41] Of course, if people obtain substantial utility as a result of workplace health protections, these can be assigned great weight in a utilitarian calculus. (Theoretically they could have an *infinite* value, but no one to my knowledge endorses such a position. Furthermore, if they had an infinite value on the utilitarian account, *all* resources would have to go to serve this end, unless some other utility-producing goal also had infinite value.) In assigning urgency or importance to health protections utilitarianism will regard them as fairly important and thus assign substantial urgency to false negatives and underregulation. It will, however—at least in most versions of the theory—permit less urgent interests experienced by enough people to outweigh the need for health protections.[42]

(2) In addition, utilitarianism permits a decrease in health care protections in return for an increase in job opportunities for any group in society (even those well off in terms of opportunities and wealth), as long as more utility is produced by increased job opportunities. There is no special attention paid to comparative distributions of social goods between social classes within the community; the theory of justice considered next differs substantially in this respect. Thus, not only do health care protections have no special place in utilitarian distributions, but there is no particular consideration for protecting those in the community who are worst off.[43] Yet this seems an important consideration, since poverty is a contributor to disease.[44] Further, utilitarian theories provide no guarantee of *equal* or any other particular distribution of health protections for all members of the community, for the proper distribution is merely one that maximizes total utility (or the net change in utility).

(3) Perhaps a more serious distributive problem is that the logic of utilitarianism permits *minor*—even trivial—benefits to many people to outweigh *severe* harms to a few. Thus, if enough people benefit in minor ways from products produced by industries that expose their employees to carcinogens, this might be judged a permissible outcome on utilitarian grounds. (Such judgments are difficult to substantiate in the absence of concrete facts, but in principle utilitarianism is open to such objections if *enough* people receive the minor benefits.) The reason for this is that if enough utility is produced by minor benefits to large numbers of people it will outweigh great disutility to a very few. How many must benefit in minor ways to outweigh serious harms to a few will of course depend upon their respective utilities. This feature of utilitarianism is captured by J. J. C. Smart, a well-known advocate:

> If it is rational for me to choose the pain of a visit to the dentist in order to prevent the pain of a toothache, why is it not rational of me to choose a pain for Jones, similar to that of my visit to the dentist, if that is the only way in which I can prevent a pain, equal to that of my toothache, for Robinson? Such situations continually occur in war, in mining, and in the fight against disease, when we may often find ourselves in the position of having in the general interest to inflict suffering on good and happy men.[45]

The chief problem revealed in this quotation is that although Jones and Robinson are acknowledged as distinct persons, the utility to each is aggregated as if they were a single person. The benefits to one person or group can be aggregated with and balanced against the injuries to another person or group no matter what the severity of those injuries (as long as it passes the net balance test). Thus, utilitarianism does not take the *separateness* of individual persons as seriously as it should and as other views do.[46] The major way in which this shortcoming manifests itself is in distributions that do not provide for the minimum health, welfare, and freedom needs of people affected by such distributions.[47] The failure to provide minimum protections is particularly difficult for those on the bottom tiers of a community who may be forced to work in the jobs most contaminated by toxic substances. To modify this example, Jones may suffer the contamination and the pain of, say byssinosis, while Robinson receives the benefits, perhaps cheaper cotton products, but *no one person* experiences both. And it is not clear why Robinson's (and many other persons) minor benefit should justify Jones's suffering the disease (unlike the case where one individual experiences both the benefit and the pain). Such examples are one main area of controversy between utilitarians and justice advocates.[48]

(4) Furthermore, underlying utilitarianism is a model of the good for human beings that is problematic. Utilitarians tend to regard the good of persons to be the maximization of their preferences or the maximization of an individual's happiness. By contrast the principle considered next is based on the idea of a person carrying out a rational plan of life that seeks to satisfy different needs and pursue different interests. For a utilitarian the standard for making interpersonal comparisons of well-being is the extent to which different individuals' preferences are satisfied. By contrast, a more plausible view in my judgment compares individuals' well-being without relying exclusively on the preferences or states of conciousness of individuals.

> The appropriate standard . . . is one which provides an indication of the extent to which different individuals' needs and interests are met, with different human needs and interests deemed to have varying degrees of urgency, given the regulative assumption that a person's good consists in successfully carrying out a plan of life. In other words . . . a standard relative to which some human interests can be judged more important than others. And an adequate measure of well-being is . . . one which these differentially important interests are met.[49]

Thus, deeper conceptions of human good underlie the different principles of distribution identified by utilitarianism and competing views.

If one believes with Rawls,[50] Scheffler, and others, that "human good is heterogeneous because the aims of the self are heterogeneous,"[51] and that different needs and interests have varying degrees of importance, then this feature of utilitarianism will be seen as a disadvantage. The satisfaction of preferences or desires does not capture well the varying degrees of urgency different "needs and interests" of persons.[52]

In sum, utilitarians have difficulty guaranteeing a certain *distribution* of resources because any distribution must contribute to maximizing overall community utility, where maximizing community welfare may be inconsistent with securing minimal protections for persons, such as health protection rights, or for certain groups of people in the community.[53]

Moreover, utilitarianism in the first place "notices or worries about" the frustration of utility—causing misery or frustrating the satisfaction of desires. Second, it calls attention to the wrongness of failing to secure the community good. It does not attend to particular distributions except as means to their ends. A particular utilitarian theory might even be extensionally equivalent, for example, to the theory of justice considered next, but this would have to be argued for relying upon the contingencies of maximizing utility. However, even if this were the case, as a matter of emphasis the distribution of health protections would not be highlighted. At this point, if there were such a utilitarian theory (of which I am not aware) this criticism becomes more of a psychological point, one of emphasis rather than a logical matter, but an important one nonetheless.

The above problems reveal implications for the utilitarian weighing of the costs of false positives and false negatives. For one thing, how important avoiding the costs of regulatory false negatives is (failing to identify carcinogens correctly) will depend upon the utilitarian value assigned to the adverse health effects of persons contracting cancer. Further, the costs of false negatives can be outweighed by "enough" benefits from other goods from such mistakes. That is, failing to detect carcinogens might well be permissible according to the utilitarian calculus as long as the community receives enough benefits, commercial and otherwise, from using the substance. How much in the way of benefits will be sufficient to outweigh such mistakes will depend upon the relative utilitarian weight assigned to, and the health costs and non–health benefits from, false negatives. Moreover, because utilitarians do not recognize different needs and interests with varying degrees of urgency, they are not committed to certain patterns of distribution. Thus, they do not have an obviously principled way of assigning costs to false negatives and false positives as does the OSH Act or the principle of justice considered next. Similar arguments apply to the costs of overregulation and underregulation.

The Daniels–Rawls Theory of Health Care Protection

Theories of justice contrast with the view just considered. These are typically sensitive to the *distribution* of benefits and burdens in the community and to the *special weight* or *urgency* that certain interests are thought to have. One such view that has been systematically developed in recent years I call the Daniels–Rawls theory of health protections, because, although the theory itself is due to Norman Daniels,[54] he grafts it onto John Rawls's comprehensive *A Theory of Justice*.

The theme of Rawls's book is that "each member of society is thought to have an inviolability founded upon justice . . . which even the welfare of everyone else cannot override."[55] This general point is much too abstract to guide our deliberations. However, it does suggest the conditional claim that if a person's health is one of the ways in which he should be inviolable, then in the context of Rawls's theory, it should not be sacrificed *simply* in order to increase the welfare of everyone else. In this respect, Rawls's theory contrasts with the utilitarian theory just discussed.

It is not easy to provide a Rawlsian argument for health care, since he does not address this issue at all; certainly such protections do not appear among the primary goods or among the liberties protected by his principles of justice. Daniels, however, recently argued that protection of people's health is part of securing people's equal opportunity to pursue the good things of life available in the community and through community institutions, such as wealth, positions of power and influence, good jobs, and recreational possibilities.

Rawls's theory of justice is designed in part for "normal, active, fully cooperating members of a society over the course of a complete life" who wish to use their income, opportunities, and rights to pursue their own conceptions of the good life for themselves.[56] But this general feature needs supplementation for health protections, since it appears that Rawls assumes no one is sick, threatened by disease or death at the hands of fellow citizens, or suffering from disease-produced handicaps.[57] Thus,

Daniels continues, a more realistic set of assumptions is needed to extend Rawls's theory to justify health protections. Daniels indicates that "health care has normal functioning as its goal: it concentrates on a specific class of obvious disadvantages [caused by disease] and tries to eliminate them."[58] Consequently, health care protections, much like education, for example, have "*great strategic importance* for [social] opportunity," which puts these particular "needs in a separate category from those basic needs we can expect people to purchase" from their own income.[59]

He then argues that health care protections have a special place in the lives of people and they should receive special protections in distributive theories. Such needs are among those that are "necessary to achieve or maintain species-typical normal functioning,"[60] that is, normal functioning as a biological member of the human species. Health care needs, Daniels claims, are those

> things we need in order to maintain, restore, or provide functional equivalents . . . to normal species functioning. They can be divided into 1) adequate nutrition, shelter, 2) sanitary, safe, unpolluted living and working conditions, 3) exercise, rest, and some other features of life-style, 4) preventive, curative, and rehabilitative personal medical services, 5) non-medical personal and social support services.[61]

Furthermore, normal species functioning is important because "impairment of normal functioning through disease and disability restricts an individual's opportunity *relative to that portion of the normal range his skills and talents would have made available to him were he healthy"*; a person's *normal opportunity range* "for a given society is the array of life plans reasonable persons in it are likely to construct for themselves."[62] What this range is in a particular society depends upon its technological development, material wealth, the kind of culture it has, and its historical development.[63] A person's normal opportunity range would be much greater in a country such as the United States with its tremendous material wealth and technological development; it would be much smaller in Bangladesh, one of the poorest countries in the world. It is also a function of a person's particular talents and abilities. Wolfgang Amadeus Mozart or John Stuart Mill will have a greater normal opportunity range than will a person affected with Down's syndrome. In ensuring equal opportunities for persons in the community, each individual's talents and abilities are not to be made *equal*. Instead opportunities must be *equal* for persons with *similar* talents and abilities.[64] Thus, the appropriate opportunity range for an individual depends in part upon his own talents and abilities and in part upon the community in which he lives.

To the extent that disease and disability undermine a person's normal opportunity range, given his talents and abilities and his particular society, he is denied fair equality of opportunity to pursue his plans of life in that community. For society as a whole

> we should use impairment of the normal opportunity range as a fairly crude measure of the relative importance of health-care needs at the macro level. In general, it will be more important to prevent, cure, or compensate for those disease conditions which involve a greater curtailment of an individual's share of the normal opportunity range.[65]

> Subsuming health-care institutions under the opportunity principle can be viewed as
> a way of keeping the system [of justice] as close as possible to the original idealiza-
> tion under which Rawls' theory was constructed, namely, that we are concerned
> with normal, fully functioning persons with a complete life span.[66]

An important set of institutions can be viewed as a first defense of the idealization: they act to minimize the likelihood of departures from the assumption that
people are capable of normal, active lives and capable of cooperating fully over the
course of a complete life. Prominent here are institutions that provide for public
health, environmental cleanliness, preventive personal medical services, occupational health and safety, food and drug protections, nutritional education, and educational and incentive measures to promote individual responsibility for healthy life
styles.

Thus, justice requires securing equal opportunities for people in a community,
and receiving equal opportunity requires the securing of these environmental health
protections. Environmental health protections, then, are a requirement of justice on
this view.

The rationale for protecting health is the same as the rationale for securing
universal education: it goes a long way toward correcting some of the inequities that
may be induced by alterable social institutions, by the (bad) luck of genetic endow-
ment, or by other adventitious events in people's lives.

> If it is important to use resources to counter the advantages in opportunity some get
> in the natural lottery, it is equally important to use resources to counter the natural
> disadvantages induced by disease (and since class-differentiated social conditions
> contribute significantly to the etiology of disease, we are reminded disease is not
> just a product of the natural component of the lottery).[67]

Health care has as its goal normal species functioning and so "concentrates on a
specific class of obvious disadvantages and tries to eliminate them. That is its limited
contribution to guaranteeing fair equality of opportunity."[68]

In the case of *human*-caused diseases resulting from contaminated workplaces,
water, air, or food, there is an additional consideration: human practices or institutions *that could be modified* produce the diseases that undermine equal social opportunities. Thus, diseases induced in the workplace or resulting from exposure to
carcinogens in the wider environment and that are caused by human activities are a
matter of injustice on two counts: they undermine opportunities in morally arbitrary
ways just as educational deficiencies do; and they are the result of *alterable* human
practices in much the same way as conscious or unconscious discriminatory practices in educational institutions are.

Finally, the theory rests on a crucial assumption: justice requires equality of
opportunity, that is, opportunity to pursue those aspects of one's life plan that one
regards as desirable, whatever they might be. Conceived narrowly, however, as
merely the equal opportunity to pursue jobs and offices that have benefits connected
to them, this is a limiting rationale for two reasons: if persons are too old to have
many opportunities open to them, then there is no longer a strong rationale on
grounds of equality of opportunity for protecting health; and it is not obvious that the
only rationale for protecting health is to secure equality of opportunities. Conse-

quently, Daniels broadens the notion of opportunity to take account of different stages in one's life: "nurturing and training in childhood and youth, pursuit of career and family in adult years, and the completion of life projects in later years."[69] Nonetheless, Daniels's rationale may not be a fully comprehensive rationale for health care protections. As Buchanan notes, "health care often relieves suffering, prevents unwanted death or enhances one's capacity for enjoying what everyone is able to do, even when it does not extend one's range of opportunities."[70]

The Daniels–Rawls theory, however, does have several advantages over a utilitarian justification for workplace health protections.

(1) It specifies a certain distribution of health protections as desirable; it aims to be *egalitarian*—to secure equal opportunities and, insofar as health protections promote equality of opportunity, to secure equal health protections. Other things equal, everyone as a matter of justice has an *equal* right to such protections.[71] The principle assigns an urgency to a normal opportunity range to each person and assigns equal health care protections insofar as this is a means to the end of equality of opportunity. This might be seen as especially important in workplace and environmental health protections, since these constitute part of the background features of community life which condition competition and cooperation in the community, as well as a person's pursuit of life goals. Such protections would include minimally and equally clean air, water, and food supplies for all members of the community. In addition, one would think that part of securing equal opportunity should include a place of work where one's life prospects would not be substantially damaged, unless it was part of one's life style, such as working as a stunt double or working on high rise buildings.[72]

(2) Once health care is secured as a matter of equality of opportunity, the importance of such protections in a society that values autonomy and individual liberty is clear. Ideally, equal opportunities are secured for all, so individuals have a range of choices commensurate with their talents and abilities. If their opportunities are greatly constrained or they have to choose between being on welfare or working only in dangerous jobs, they do not live in a minimally just society. If they choose to work in dangerous jobs when they have little choice, they have not justly consented to them. Moreover, if in fact they have a normal range of opportunities, then the idea that they legitimately consent to the choices they have made has much greater validity.

(3) Since Daniels embeds his theory within Rawls's theory of justice, with its stringent ordering principles, it is not permissible to trade away equality of opportunity merely for an increase in the general welfare, for an increase in income either to companies or to their shareholders, for a decrease in the cost of products to consumers, or for increased income or wealth for the general public.[73] Thus, the Daniels–Rawls theory provides somewhat more precise guidance regarding the health–wealth tradeoff than does utilitarianism, and it gives a correspondingly greater security to workplace health protections. Whether Rawls's lexical ordering principles are defensible is an issue we cannot pursue further here. In any case many others have evaluated it.[74]

(4) From paragraph 3 it follows that *minor* benefits to *many* cannot override one's health protections secured by the equal opportunity principle, even if doing so

maximizes the general welfare or the gross national product. Some things are more important than producing net community benefits. In particular, securing fair opportunities for individuals to plan and pursue their conceptions of the good life for themselves free from the adventitious losses posed by diseases produced by the alterable activities of other human beings is more important than net increases in social welfare.

Health protections have a special weight compared to the general welfare. To paraphrase the quotation earlier:

> It is not rational, or at least not *just,* for me to choose to damage Jones's health (if it reduces his normal opportunity range), even if this will increase Robinson's and Smith's income. Jones's health is inviolable vis-à-vis their wealth in such circumstances.

Thus, Daniels's principle does two things identified here and in the preceding paragraph. It implicitly assigns a certain importance or urgency to health protections, and thus an implicit urgency to avoiding regulatory false negatives. It also indicates that the interest in health protections according to his principles of justice cannot be easily outweighed by minor benefits to others.

(5) Thus, the *distributive* considerations central to this view constrain how resources are allocated. For example, other things being equal, a firm may not contaminate groundwater with carcinogens if this will threaten human drinking water, even if the firm can lower the costs of doing business and help the economy by disposing of its waste products in this way. Or a firm may not expose its employees to disease- and death-causing substances if this will interfere with their normal opportunity range to pursue their plans of life.[75]

One drawback to this moral view is that it provides only the rough guidance indicated previously (curtailment of persons' normal opportunity range at the macro level) concerning the *extent* of health protections for environmentally caused diseases, especially the rarer ones typically caused by carcinogens. In the United States there is a lively debate about how clean the water, air, workplace, and food should be, and even how serious these threats to health are compared with other environmental risks, such as, bacterially or virally carried diseases in impure drinking water. Must all risks from environmentally caused diseases be eliminated, or must they be reduced so that there is less than a 1 in 1,000,000 chance of contracting a disease from lifetime exposure to a toxin? The theory stops short of addressing these debates.

(6) With regard to the tradeoff between workplace health protections and increasing opportunities, the Daniels–Rawls principle provides more precise guidance than does utilitarianism. According to Rawls, if, because of scarcity, one cannot secure equality of opportunity for all, then one may increase opportunities for those with the fewest opportunities at the expense of not having as much equality of opportunity. As he puts it, "We must . . . claim that the attempt to eliminate . . . inequalities [of opportunity] could so interfere with the social system and the operations of the economy that in the long run the opportunities of the disadvantaged would be even more limited."[76] We "must hold that a wider range of more desirable alternatives is open to them than otherwise would be the case."[77]

In the context of workplace health protections, this suggests that if one can increase total job and office opportunities for those who have the fewest by decreasing (or by not providing) health care protections for them, then even Rawls's special conception of justice permits, perhaps even requires, such a decrement. The principle appears to permit a decrease in (or failure to increase) health care protections to the extent that this increases opportunities for those with fewest opportunities. It also permits this decrease (or failure to increase) to the extent that it maximizes job opportunities for those who are worst off in this regard. Thus, the theory permits the health care protection–job opportunities tradeoff, as does utilitarianism, but for more limited reasons. However, this is potentially a substantial shortcoming which defenders of the theory should address, for depending upon factual matters it appears to permit a retreat from health protections in order to increase opportunities for those who are poorest (and probably in the environmentally "dirtiest" jobs) or who live in the environmentally "dirtiest" areas.

Despite some obvious advantages of the Daniels–Rawls theory, there are several shortcomings. First, the definition of opportunity is not entirely clear. Ideally, one would like to be able to classify opportunities so that one could individuate and count them, so that the idea of maximizing them would make sense. In this presentation, the best we can hope for is to use our intuitive notion of an opportunity and a similar intuitive notion of increasing and maximizing opportunities.

Second, because health care protections are lumped under the equality of opportunity principle, it is not clear under conditions of scarcity how one would trade off spending on education or health care protections in the workplace, both of which are part of securing equality of opportunity. Which distribution of resources will do more to secure the appropriate opportunities for those in the community? Both of these objections point to the need for greater specificity in the theory.

Third, as noted, this theory still leaves somewhat vague the tradeoff between securing health care protections and increasing opportunities. Would the Daniels–Rawls theory permit the cotton industry to remain "dirty" (i.e., to cause a higher incidence of byssinosis among those working in the cotton industry compared with the general population) if this would increase the job opportunities for the poorest strata of population living in especially poor parts of North and South Carolina? Or does proper application of the principle require us to look at the impact of legislation such as the Occupational Safety and Health Act across the entire country and not industry by industry and disease by disease? What is the appropriate worst off group of concern according to the theory?

Fourth, and more seriously, as we can see from the preceding point, there is an important respect in which the Daniels–Rawls theory may be worse than the Occupational Safety and Health Act in providing workplace health care protections in order to help make opportunities for the good life equal: it does not indicate as precise an absolute level of health care as does the OSH Act. It is a *comparative* principle, seeking to equalize opportunities between persons, whereas the principle implicit in the OSH Act is a noncomparative one that focuses (at least in theory) on the well-being of particular individuals. On Daniels's view workplace health protections may always be traded for increasing opportunities, but the idea of "increasing opportunities" even for those worst off in some circumstances makes it difficult even on Rawls's principles to justify increasing workplace health protections. This tradeoff is potentially a serious objection to the theory because it appears that equal opportunities and the health protections they provide might be frequently sacrificed to increase opportunities for the worst off. The Occupational Safety and Health Act, by contrast, tells us to see to it that *no* person suffers *material impairment of health or functional capacity* even if such person be exposed to a potentially toxic substance for the duration of his working life. As long as we have an adequate conception of health and what counts as "material impairment," we have a fairly accurate goal for which to strive that is not quite as comparative as Daniels's suggestions.

The "no material impairment" principle has a normative status something like a right, which appears to trump general social goods, until workplace health protections become economically "infeasible." In ideal circumstances the OSH Act and Daniels's principle should approximate each other since "material impairment of health or functional capacity" is likely to decrease a person's "normal opportunity range." However, in less than ideal circumstances it appears that Daniels's theory might allocate fewer resources to workplace and environmental health protections than would the OSH Act. This is not necessarily to argue that the OSH Act is superior, for its uncompromising goal of health protections may not be realistic in many circumstances, and that theoretical goal is substantially compromised by the technological and economic feasibility requirement and by evidentiary requirements.

Although the preceding discussion does not solve the problem of the justice of workplace health protections, it does provide some of the main outlines of an adequate theory. (1) We must have a better rationale for protecting persons from workplace and environmental contaminants; both utilitarians and Daniels have part of the truth on this matter. An adequate rationale would seek to prevent the suffering of persons from human environmentally caused diseases, to prevent their suffering diminished capacity for enjoying what everyone else is able to do, and to secure equality of normal opportunity ranges for them (consistent with their skills and abilities) which workplaces and environments contaminated with carcinogens threaten. (2) Next, we need a relatively precise principle to guide the health–wealth tradeoff. (3) We need a relatively precise principle to guide the health–opportunity tradeoff. The Daniels–Rawls theory provides some guidance but still does not satisfactorily solve the problem of employees trading away their good health simply to have more job opportunities rather than fewer or simply in order to have some rather than none, for the idea of maximizing opportunities is a greedy concept. It seems one could almost always increase opportunities, even for the worst off group in the community. The idea is to provide the worst off economic class with opportunities up to the point at which their total opportunity set will be *maximized*. It appears difficult to know exactly at what point this will be reached and how to make this idea more precise.

(4) Finally, it is preferable for purposes of precision and better for guidance, but not necessary, that one have a noncomparative or less comparative standard of health protection to guide workplace health protections. The reason for this is that one can achieve equality of opportunities (Daniels's rationale) by providing few opportunities for each—by securing equal, but poor, health care protections for all.

What is needed instead is something more like the requirement of the Occupational Safety and Health Act, that there be no material impairment of health, to ensure that in most circumstances a person will not have to choose between having a job or having good health. This last consideration is not necessary for reasons we have just seen, and because rights to social goods pose their own set of problems—they frequently can trump too many other things. However, points 1–4 provide the outlines of desirable features of a theory.

I believe a number of theories, including Daniels's (suitably modified), might satisfy these conditions, even including some consequentialist theories suggested by Scanlon and Scheffler.[78] In the last section I indicate the attractions of a theory of justice or of a "distributively sensitive" consequentialism for ensuring workplace and environmental health protectors.

The Attraction of Distributively Sensitive Theories

Throughout I have been concerned about the possibility of false negatives, under-regulation, and uncompensation resulting from evidentiary procedures in the law. These concerns appear to have received too little attention in the tort law and in administrative agencies. Yet on most theories of distribution, even including utilitarianism, there would be considerable weight assigned to workplace and environmental health protections. (Although, as we have seen, utilitarianism has greater difficulty assigning prominence to these concerns than does the OSH Act or the Daniels–Rawls view.) Thus, there should be considerable urgency to avoid the epistemic mistakes represented by false negatives and for underregulation.

The moral theories that seem most defensible provide some account for this concern. The Daniels–Rawls theory possesses many elements of such a defensible view. I want to build on the insights of that theory to motivate, at least in outline, workplace and environmental health care protections and the substantial costs of regulatory false negatives. I develop this view partly by contrast with utilitarian theories for two reasons: *most* forms of utilitarianism exhibit many of the principal shortcomings I wish to avoid; and utilitarianism and some of its offshoots, such as economic cost–benefit analysis, have tended to dominate the policy discussions on environmental health protections.[79]

The key to my concerns lies in the *distributive* features of moral theories; utilitarian theories do not provide this except as a means to maximizing utility, whereas justice theories tend to. There are utilitarian strategies for meeting this concern: providing special weights for certain preferences people have (e.g., special weight for the desire for good health), emphasizing the law of diminishing marginal utilities, and so on. However, even when these strategies are adopted, most utilitarian theories still aggregate the individual utilities and choose among them so as to maximize the result, thus treating the community much like a single person. Minor benefits for many, therefore, can take precedence over severe harms to a few because each person's wants or, more important, each person's *needs*, including health care needs, are seen as just another way of producing community utility. Yet some of these needs, good health in particular, have a special role or urgency in the life of each person and are important in evaluating the relations between persons in the community that appears not to be captured by the typical utilitarian treatment.

An attraction to a theory which assigns urgency to environmental and workplace health protections is that it selects the *distribution* of these goods to notice and worry about. By identifying health protections to be of concern, it calls attention to them and makes them a less contingent matter than utilitarian principles might.

We can begin to understand some of these points by considering the case of Steve Talbert described in the letter from his wife, which is the epigraph to this chapter. Talbert was exposed to so much cotton dust that he contracted byssinosis. His disabilities were so bad that he had to quit his job, give up his favorite recreational activities, which very likely were among the things that made his life worth living, and sell many possessions (including his now useless recreational equipment) to pay his high medical bills. Finally, he became tethered to an oxygen tank 24 hours a day.

Of course, most normative ethical theories will condemn Steve Talbert's treatment as wrong. What utilitarian and other maximizing theories underemphasize or fail to call sufficient attention to is the strategic importance and special urgency of good health for the Steve Talberts of the world. His disease was disastrous for him and for his family—his wife's letter conveys this. Of course he was miserable, suffered much, and was prevented from having very important wants satisfied, as the utilitarians would argue. More importantly, his "normal opportunity range" for enjoying life, for supporting his family, for completing his life plans, and for doing almost anything that makes life worth living was destroyed. Daniels's view comes closer to capturing what is wrong about not protecting Steve Talbert from such harms, for Talbert's life has quite literally been ruined by his workplace-contracted disease.

However, not only has he lost something of almost incalculable importance to him and his family because of his disease, but the circumstances under which he contracted his disease also provide a clue to the wrongness of the relationships that led to his injuries. It is as if something of great importance (his good health) has been taken from him and converted to the benefit of others. Of course others benefited because insufficient resources were spent to protect him from byssinosis. The cotton mill had somewhat lower production costs, its goods were priced lower as a consequence, and it perhaps had higher profits as well. Indeed many of us may have benefited in very minor ways from his and other mill workers' illnesses by purchasing somewhat lower priced cotton goods. However, these benefits as they affect each individual tend to be minor in magnitude compared with the losses Talbert suffered. Recall that this is typical of environmental risk problems: toxic substances typically have catastrophic effects on their victims, but the benefits from their use tend to be relatively modest by comparison.[80] In short, he seems to have been treated unfairly in suffering his loss: the *relationship* which led to his disease was wrong. It appears that there was a *maldistribution* of benefits and harms in this relationship.

Now it is not that utilitarian theories would not condemn this relationship as wrong; no doubt all, or nearly all, would. The problem is that they fail to call sufficient and appropriate attention to different individuals varying needs and interests and to the distribution of social goods and harms in the community. They appear not to assign sufficient urgency to health harms vis-à-vis other goods in the community. For all its virtues, utilitarianism's most serious shortcoming for distributing the costs and benefits of toxic substances is that it permits severe, even catastrophic, harms to some to be outweighed by minor benefits to many. This defect is well known, yet utilitarianism is a view to which many gravitate for justifying the use of toxic substances—perhaps precisely because of this feature. Thus, and this is slightly vague, the properties typical of toxic substances—providing minor benefits to many but potentially severe costs to a few—appear to be nicely permitted by utilitarian views. It may be no accident that such views have been prominent in arguments in the environmental health area. If this is right, we also need a paradigm shift in how we think about the morality of environmental health protections: we should have much more distributively sensitive views.

One virtue of justice theories such as the Daniels–Rawls view or of a distributively sensitive consequentialism of the kind suggested by Scanlon or Scheffler is that they call attention to the distributions of goods that make a substantial difference in the lives of individuals and their families. They also provide a way of analyzing what is wrong with relationships that lead to such serious harms for the Steve Talberts of the world. And they call attention to the particular urgency which certain needs (e.g., health care needs) should have.

In presenting these views I have not argued for a particular distributive theory. In fact it seems that probably several different consequentialist and nonconsequentialist views might well provide appropriate guidance on these matters as long as they measure up to the somewhat vague guidelines set out here. Clearly much more needs to be said on these issues (and I have not even ruled out some views that I might find mistaken). I have tried, however, to sketch enough of such theories to indicate the importance of some of the concerns expressed throughout the book.[81]

I believe in fact that a distributively sensitive consequentialism has much to recommend it, although I do not argue at length for this. I have just rehearsed some of the attractions of *distributive* concerns. *Consequentialism* is attractive for several reasons. The future consequences of an act or policy are morally relevant considerations in a way that they are not always for nonconsequentialist views. Thus, some of the knowledge-generating and cost-avoidance effects of tort law principles (discussed in Chapter 3) would be supported by consequentialism. In addition, on a consequentialist view the right thing to do is to produce the best outcome. This permits one to take into account the overall effects of one's actions in addition to distributions. This may permit efficiency considerations to enter into justification of a course of action, a definite attraction for principles guiding public policy decisions. Clearly, these considerations do not provide a justification for a distributively sensitive consequentialism; they merely call attention to attractive features. Much more needs to be done by way of evaluating and justifying them.[82]

In articulating what I see as the wrong done to Steve Talbert, I avoided reliance on the idea (and it may not be true) that his employers were consciously using or exploiting him. Indeed one aim of the arguments of this book is to call attention to institutional approaches to assessing the risks from toxic substances that are for the most part not consciously designed to use, exploit, or steal from employees, the general public, or plaintiffs in tort cases. By calling attention to different approaches to the ''science'' of assessing the risks from toxic substances, I hope to prevent our evidence-gathering activities from *inadvertently* producing results that damage the

Steve Talberts of the world. I have tried to provide reasons for approaching the scientific evaluation of evidence in the tort law and administrative regulations so that these procedures do not frustrate one of the major aims for engaging in the activity in question—to discover whether there exist health harms to the public, a concern that has special urgency for the public and to which distributively sensitive theories will assign considerable moral weight. In the tort law we should not be so concerned to have evidence of harm from toxic torts that is scientifically defensible in the best journals that the wrongful loss of a person's good health goes uncompensated. In administrative agencies we should not be so concerned to develop biologically correct models for assessing each risk in question that we identify few carcinogens or assess few of the risks of those we have identified.

We should not be so concerned about the fine points of scientific inference that we lose sight of the effect of our activities on the lives of individuals and on the relationships in which we all live. Scientific inquiries should be designed in part to serve the moral and political goals of the institutions in which they are used. This has not always been the case; it is not what many experts in those fields are currently recommending; but it seems the correct approach in order to secure a better distribution of health care protections in the community.

In the present state of knowledge, and perhaps for a long time to come, we are probably condemned to regulating toxic substances "through a glass darkly." In such circumstances it seems even more important to have a clear moral view about the effects of our evidentiary procedures on the lives of persons in our communities. That view, I have suggested, is a theory of justice or a distributively sensitive consequentialist or nonconsequentialist theory, one sensitive to the distribution of health care protections. This will place greater urgency on avoiding false negatives and strike a better balance between false positives and false negatives than present practices appear to. Such guidance will not clarify the glass through which we acquire our information. It will provide needed guidance under conditions of uncertainty. It might even help ensure that our evidentiary procedures do not inadvertently frustrate many of the health-protective goals of our institutions. A more optimistic view is that a better understanding of distributively sensitive theories and a corresponding critical view of utilitarianism will remind us what is at stake when we have to make decisions about possible carcinogens under conditions of considerable uncertainty and may lead us to adopt scientific procedures in tort and administrative law that will more readily secure health care protections.

Notes

INTRODUCTION

1. In discussing toxic substances, I focus on carcinogens. There has been more study of and experience with carcinogen risk assessment because carcinogens have been the object of substantial regulatory activity. The scientific models for characterizing the effects of carcinogens on human beings seem better developed than, for example, those for reproductive toxins or neurotoxins. In addition, the debates about the scientific models and information seem more manageable and confined, and the disputes between different scientific strategies more clearly marked out than for other kinds of toxins. Finally, I am more familiar with the numerous scientific and regulatory issues concerning the regulation of carcinogens.

In 1985–86, while serving as a congressional fellow at the U.S. Congress's Office of Technology Assessment, I was coauthor of the report *Identifying and Regulating Carcinogens* (Washington, D.C.: U.S. Government Printing Office, 1987). I have since served as the principal investigator on the University of California, Riverside, Carcinogen Risk Assessment Project funded first by a grant from the University of California Toxic Substances Research and Teaching Program ("The UCR Carcinogen Risk Assessment Project") and then by National Science Foundation Grant No. DIR-8912780 ("Evidentiary Procedures for Carcinogen Risk Assessment").

2. Sanford E. Gaines, "A Taxonomy of Toxic Torts," *Toxics Law Reporter,* 30 November 1988, p. 831.

3. Talbot Page, "A Generic View of Toxic Chemicals and Similar Risks," *Ecology Law Quarterly* 7 (1978):208.

4. Ibid., p. 209.

5. Ibid., pp. 210–11. Note that they have other properties as well. Benefits from using these substances are typically internalized through markets and product prices while the costs are imposed on others through nonmarket mechanisms, creating externalities. Finally, many impose collective risks that materialize only after a substantial latency period (pp. 212–213).

6. U.S. Congress, Office of Technology Assessment, *Cancer Risk: Assessing and Reducing the Dangers in Our Society* (Boulder, Colo: Westview Press, 1982), pp. 70–71.

7. Ibid., pp. 69–72.

8. The *New York Times* reported that secondary smoke causes both cancer and coronary problems. Lawrence K. Altman, "The Evidence Mounts Against Passive Smoking," *New York Times,* 29 May 1990, p. 135. More recent information reviewed by an Environmental Protection Agency Science Advisory Board found that environmental tobacco smoke caused about 3700 excess cancers per year. *Environmental Health Letter* 30 (9), 23 April 1991, p. 92.

9. U.S. Cong., *Cancer Risk,* pp. 86–88, 91, 108.

10. K. Bridbord and J. French, *Estimates of the Fraction of Cancer in the United States Related to Occupation Exposures* (National Cancer Institute, National Institute of Environmental Health Sciences, and National Institute for Occupational Safety and Health, 15 September 1978).

11. Quoted in Samuel S. Epstein and Joel B. Swartz, "Fallacies of Lifestyle Theories," *Nature* 289 (1981):130.

12. Ibid.

13. Ibid., pp. 127–28.

14. Ibid., pp. 127–30, and D. Schmahl, R. Preussman, and M. R. Berger, "Causes of Cancer—An Alternative View to Doll and Peto," *Klinische Wochenschrift* 67 (1981):1169–73.

15. For a discussion of these issues see L. Tomatis, A. Aitio, N. E. Day, E. Heseltine, J. Kaldor, A. B. Millers, D. M. Paskin, and E. Bibali, eds., *Cancer: Causes, Occurrence and Control* (Lyon, France: International Agency for Research on Cancer, 1990), L. Tomatis, "Environmental Cancer Risk Factors: A Review," *Acta Oncologica* 27 (1988):465–72, and Schmahl, Preussman, and Berger, "Causes of Cancer."

16. A recent report estimates that 5–10% of all cancer deaths are caused by workplace conditions alone and, as we have seen, the percentage may be much higher. Johnny Scott, "Job-Related Illness Called 'America's Invisible Killer,'" *Los Angeles Times,* 31 August 1990, p. A4.

17. As OTA put it, this does not "represent a decision that chemicals, in the workplace or in the general environment, are more important in cancer causation than dietary elements, personal habits, radiation or certain aspects of human biology. However, it does reflect the major legislative and regulatory emphasis recently placed on chemicals, and the greater ease with which chemical carcinogens can be detected by present-day methods." U.S. Cong., OTA, *Cancer Risk,* p. 12.

18. Ibid., pp. 12, 130.

19. These results probably exaggerate the number of positives and the "relaxed criterion" of carcinogenicity would probably "not be accepted in any regulatory decision." Ibid.

20. Ibid. The Occupational Safety and Health Administration had a contractor examine a list of 2400 suspected carcinogens compiled by the National Institutes of Health and that agent estimated for about 24% of the total (570) there were sufficient data to classify them as carcinogens under OSHA's policies. More up to date information was compiled by the National Cancer Institute (NCI), the International Agency for Research on Cancer (IARC), and the National Toxicology Program (NTP). "Fifty two percent of NCI-tested and reported chemicals were carcinogens. Either 'sufficient' or 'limited' data existed to classify about 65 percent of IARC-listed chemicals as carcinogens . . . [and] NTP . . . reported that [of] 252 . . . tests . . . completed in the NCI Bioassay Program 42 percent were positive, 9.5 percent were equivocal, 36 were negative and 13 percent were inconclusive."

The U.S. Congress's Office of Technology Assessment reports that tests by the NTP show that of 308 substances tested 144 were carcinogenic in one or more animal species. Sixty-one substances indicating much greater potency were positive in three or four animal tests. *Identifying and Regulating Carcinogens,* pp. 18–21. Somewhat more recently the NTP reports that of 327 studies 159 (48%) were carcinogenic in one or more animal sex-species. J. K. Haseman, J. E.Huff, E. Zeiger, and E. E. McConnell, "Comparative Results of 327 Chemical Carcinogenicity Studies," *Environmental Health Perspectives* 74:229.

It once appeared that there might be a bias toward "likely-to-be-positive" chemicals. And negative tests were less likely to be published. Both facts might inflate the numbers of carcinogens. However, it is not clear how much of a bias there is in choice of substances for testing. Recent testimony by Jack Moore, then acting assistant administrator for pesticides and toxic substances for the Environmental Protection Agency, before California's Proposition 65 Science Advisory Panel in April 1987 indicated that the NTP substances were tested primarily because of their high production volume, not necessarily their suspect toxicity. Moreover, anecdotal evidence now indicates that negative tests are more likely to be reported.

21. Even though some 6000–7000 substances may have been evaluated in animal studies,

a data base of such studies compiled by Gold et al., which screens them carefully for good studies, contains more than 4000 studies, but they were performed on only 1050 chemicals. California Environmental Protection Agency, Health Hazard Assessment Division, "Expedited Method for Estimating Cancer Potency: Use of Tabulated TD_{50} Values of Gold et al.," Presented to Proposition 65 Science Advisory Panel, 26 April 1991.

22. Few benzidine dyes have been regulated, yet researchers have found a number of benzidine-related compounds that appear to be as problematic as regulated dyes. William M. Johnson and William D. Parnes, "Beta-Naphthalamine and Benzidine: Identification of Groups at High Risk of Cancer," in *Annals of the New York Academy of Sciences: Public Control of Environmental Health Hazards,* ed. E. Cuyler Hammond and Irving J. Selikoff (New York: New York Academy of Sciences, 1979), pp. 277–84.

23. The State of California has identified 369 carcinogens, but as of fall 1990 only 48 of these (13%) had been evaluated in a risk assessment. The U.S. EPA apparently had performed risk assessments on about the same number of substances. As a member of California's Proposition 65 Science Advisory panel I have access to this information. Also, personal communication from Lauren Zeise, acting chief, Reproductive and Cancer Hazard Assessment Section, Office of Environmental Health Hazard Assessment, California Environmental Protection Agency.

24. Few substances that have been found to be carcinogenic have been regulated. For example, OSHA has regulated only 22 substances in 15 years. The Environmental Protection Agency under the authority of the Clean Air Act as of November 1987 had regulated only 4 of 37 toxic substances identified in 1977 for special consideration and required by Congress to be considered for regulation. Even when there are substances of known carcinogenicity, few have been regulated fully. As of November 1987, the National Toxicology Program has identified 144 substances as carcinogenic in animal bioassays, but no agency has regulated more than one-third of these and no regulatory action at all has occurred for 43% of the 144 substances. U.S. Cong., OTA, *Identifying and Regulating Carcinogens,* p. 19.

25. Toxic substances would be controlled by means of the criminal law, if it were made a crime, punishable by jail terms or fines, to harm others or the environment by the release of toxic substances. Typically the criminal law in addressing these issues would require that the perpetrator exhibit the appropriate "guilty mind" by acting intentionally, knowingly, recklessly, or negligently in committing the offense. Although I do not discuss the criminal law control of toxic substances, some high-visibility criminal cases have been brought. See "Job-Related Murder Convictions of Three Executives Are Overturned," *New York Times,* 20 January 1990, p. 10, for a report on one such case. See C. Cranor, "Political Philosophy, Morality and the Law," *Yale Law Journal* 95 (1986):1066 for some discussion of the use of criminal and contract law for the regulation of toxic substances.

Toxic substances might be controlled by means of the contract law, for example, if a group of people (say, a union) contracted with someone who might expose them to toxic substances (say, the company for which they worked) not to expose them or be in breach of contract. The workplace is an obvious setting in which this legal approach might be taken.

26. Tort law control of toxic substances, although seemingly not the most effective, as I will argue, appears to have some deterrent effect. C. Eads and P. Reuter, *Designing Safer Products: Corporate Responses to Product Liability and Regulation* (Santa Monica, Calif.: Rand Corporation, 1983), note that nine large American manufacturers "found that products liability law administered by the courts, though generating an 'indistinct signal,' affects manufacturers' safety decisions far more than do market forces or the prospect of regulation" (p. 1087).

27. The Food, Drug and Cosmetic Act, 1906, 1938, amended 1958, 1960, 1962, 1968. 21 U.S.C. §§ 301–94. The Clean Air Act, 1970 amended 1974, 1977. 42 U.S.C. §§ 7401–7671q. The Clean Water Act, 1972 amended 1977, 1978, 1987. 33 U.S.C. §§ 1251–1387.

The Occupational Safety Health Act, 1970. 29 U.S.C. §§ 651–78. The Safe Drinking Water Act, 1976 amended 1986. 42 U.S.C. §§ 300f 300j–26.

28. The Resource Conservation and Recovery Act, 1976. 42 U.S.C. §§ 6901–92k. The Toxic Substances Control Act, 1976. 15 U.S.C. §§ 2601–71.

29. The Comprehensive Environmental Response, Compensation and Liability Act, 1980 amended 1986. 42 U.S.C. §§ 9601–75.

30. I use the terms "administrative" and "regulatory" law interchangeably.

31. See Joel Feinberg, "The Expressive Function of Punishment," in *Doing and Deserving,* ed. J. Feinberg (Princeton, N.J.: Princeton University Press, 1970), pp. 95–118, for discussion of these features of punishment.

32. A "homey" example of such prospective regulation is local building codes. If we build or remodel a house, we not only must have plans approved in advance, but a building inspector must approve of each major step of the building to ensure that it is "up to code." If a building collapses later, the parties involved, including the inspector, may then be subject to administrative as well as tort and criminal law penalties.

33. See, for example, the Occupational Safety and Health Act (OSH Act), § 6(b)(5), which requires OSHA to issue regulations for controlling exposure to toxic substances which "adequately assure" that "no employee will suffer material impairment of health or functional capacity even if such employee has regular exposure to the hazard dealt with by such standard for the period of his working life."

34. I am also more familiar with the legislation, regulation, and litigation concerning OSHA than that of some other agencies.

35. See Carl Cranor, "Collective and Individual Duties to Protect the Environment," *Journal of Applied Philosophy* 2 (1985):243–59 for a discussion of some of these issues.

36. *Industrial Union Department, AFL-CIO v. American Petroleum Institute,* 448 U.S. 607 (1980). Hereafter I refer to this as the *Benzene* case.

37. See C. Cranor, "Epidemiology and Procedural Protections for Workplace Health, in the Aftermath of the Benzene Case," *Industrial Relations Law Journal* 5 (1983):372–400 for references to some of the issues.

38. As we will see this does not mean all affected parties will be better served, but taking all social consequences into account, it appears we would be better off working under different paradigms.

39. David Rosenberg, "The Dusting of America, A Story of Asbestos—Carnage, Cover-Up and Litigation," *Harvard Law Review* 99 (1986):1693, 1704, describes the effects of asbestos litigation on the Manville Corporation, which filed for bankruptcy protection because of the potential litigation losses.

40. I believe a detailed review of the work by John Austin, Jeremy Bentham, and John Stewart Mill, to name a few, would show that their primary emphasis was the criminal law or perhaps property law. A major contribution by H. L. A. Hart was to focus more attention on the private law.

41. For example, Jeffrie G. Murphy and Jules L. Coleman, *The Philosophy of Law: An Introduction to Jurisprudence* (Totowa, N.J.: Rowman & Allenheld, 1984), extend legal philosophy discussions into these areas.

42. *Ferebee v. Chevron Chemical Co.,* 736 F.2d 1529 (D.C. Cir. 1984).

CHAPTER 1

1. I am indebted to Deborah Mayo, D. V. Gokhale, William Kemple, and Kenneth Dickey for comments and criticisms on the ideas presented in this chapter.

2. For example, in *Richardson by Richardson v. Richardson-Merrell, Inc.*, 857 F.2d. 823 (D.C. Cir. 1989), the court held that animal studies together with chemical and *in vitro* studies were not sufficient to establish a conclusion that Bendectin caused birth defects.

3. The extent to which such approaches are followed by agencies is considered in Chapter 4, but some, such as the Occupational Safety and Health Administration, the Consumer Product Safety Commission, and the Food and Drug Administration, may approach this.

4. National Research Council. *Risk Assessment in the Federal Government: Managing the Process* (Washington, D.C.: GPO, 1983), pp. 18–19. I use *risk assessment* "to mean the characterization of the potential adverse health effects of human exposures to environmental hazards. Risk assessments include several elements: description of the potential adverse health effects based on an evaluation of results of epidemiological, clinical, toxicologic, and environmental research; extrapolation from those results to predict the type and estimate the extent of health effects in humans under given conditions of exposure; judgments as to the number and characteristics of persons exposed at various intensities and durations; and summary judgments on the existence and overall magnitude of the public health problem. Risk assessment also includes characterization of the uncertainties inherent in the process of inferring risk. The committee uses the term *risk management* to describe the process of evaluating alternative regulatory actions and selecting among them. Risk management, which is carried out by regulatory agencies under various legislative mandates, is an agency decision-making process that entails consideration of political, social, economic, and engineering information with risk-related information to develop, analyze, and compare regulatory options and to select the appropriate regulatory response to a potential chronic health hazard. The selection process necessarily requires the use of value judgments on such issues as the acceptability of risk and the reasonableness of the costs of control.

5. Ibid., p. 19.

6. Ibid.

7. Even if scientists were to know exact numbers of deaths and diseases, they might be unable to determine *which* persons would be affected. Thus, even if we had a *harm assessment* we might only be able to assign probabilities to *which* individuals would contact the diseases. This point is owed to Kenneth Dickey.

8. Nicholas Rescher, *Risk: A Philosophical Introduction to the Theory of Risk Evaluation and Management* (Washington, D.C.: University Press of America, 1983), p. 5.

9. One can show this point by using actual experimental data in the computer program *Toxrisk*, which operates on a personal computer and generates risk estimates using different assumptions. See also R. C. Cothern, W. L. Coniglio, and W. L. Marcus, "Estimating Risks to Health," *Environmental Science and Technology* 20(2) (1986):111–16, who also discuss this point.

10. David Freedman and Hans Zeisel, "From Mouse to Man: The Quantitative Assessment of Cancer Risks," *Statistical Science* 3(1988):3–56.

11. It may be that the mechanism of carcinogenesis, although far from perfectly understood, is well enough understood to permit scientists doing risk assessments to rely upon the linearized multistage model, or perhaps a slight generalization of this, for making the high-dose to low-dose extrapolations at least for most substances. (Dr. Page Painter, Office of Environmental Health Hazard Assessment, California Environmental Protection Agency, personal communication.) Moreover, there also appear to be some empirical constraints on the mathematical possibilities (discussed below).

12. U.S. Department of Labor, Occupational Safety and Health Administration, "Identification, Classification and Regulation of Potential Occupational Carcinogens," *Federal Register* 45 (22 January 1980):5061.

13. U.S. Interagency Staff Group on Carcinogens, "Chemical Carcinogens: A Review of the Science and Its Associated Principles," *Environmental Health Perspectives* 67 (1986): 274.

14. D. P. Rall, "The Role of Laboratory Animal Studies in Estimating Carcinogenic Risks for Man," *Carcinogenic Risks: Strategies for Intervention* (Lyon, France: International Agency for Research on Cancer, 1979), p. 179.

15. U.S. Interagency Staff Group, "Chemical Carcinogens," quoting the International Agency on Cancer, p. 234.

16. They wanted studies of each kind that provided reasonably strong evidence of carcinogenicity in animals and humans and for which suitable data could be available for quantifying the carcinogenic potency of substances based on each kind of study. Bruce C. Allen, Kenneth S. Crump, and Annette M. Shipp, "Correlations Between Carcinogenic Potency of Chemicals in Animals and Humans," *Risk Analysis,* December 1988, pp. 531–44.

17. L. Tomatis, "Environmental Cancer Risk Factors: A Review," *Acta Oncologica* 27 (1988):465–72.

18. Asbestos is one substance that leaves a unique trace behind, as often fibers of asbestos can be found in the tissues that are diseased. Other sources of disease may be easily detected because they cause extremely rare diseases, thus raising the odds of detection (as is the case with vinyl chloride causing a very rare form of liver cancer or with diethylstilbestrol, which caused a very rare form of cervical cancer in the daughters of mothers who took DES during pregnancy).

19. Testimony of Dr. David Rall, Director, National Institute of Environmental Health Sciences, *Federal Register* 45 (1980):5061.

20. Adam M. Finkel, *Confronting Uncertainty in Risk Management: A Report* (Washington, D.C.: Center for Risk Management, Resources for the Future, 1990), pp. 12–15.

21. This point was suggested to me by Lauren Zeise, senior toxicologist and acting chief of the Reproductive and Cancer Hazard Assessment Section, Office of Environmental Health Hazard Assessment, California Environmental Protection Agency.

22. Office of the President, *Regulatory Program of the United States Government* (Washington, D.C.: GPO, 1990), pp. 13–26. See also *OMB v. the Agencies: The Future of Cancer Risk Assessment* (Cambridge, Mass: Harvard School of Public Health, Center for Risk Analysis, 1991), for a survey of some of these issues.

23. U.S. Interagency Staff Group on Carcinogens, "Chemical Carcinogens."

24. U.S. Congress, Office of Technology Assessment, *Identifying and Regulating Carcinogens* (Washington, D.C.: GPO, 1987), p. 39.

25. See John Van Ryzin, "Current Topics in Biostatistics and Epidemiology" (Discussion), *Biometrics* 28 (1982):130–38, esp. p. 135; J. Cornfield, K. Roi, and J. Van Ryzin, "Procedures for Assessing Risk at Low Levels of Exposure," *Archives of Toxicology Supplement* 3 (1980): 295–303; Lauren Zeise, Richard Wilson, and Edmund A. C. Crouch, "Dose–Response Relationships for Carcinogens: A Review," *Environmental Health Perspectives* 73 (1987): 259–308, esp. p. 263.

26. See the U.S. Department of Labor, Occupational Safety and Health Administration (OSHA), "Identification, Classification and Regulation of Potential Occupational Carcinogens," *Federal Register* 45 (22 January 1980):5190 for a discussion of different rationales for the extrapolation models mentioned here and illustrated in Figures 1-4 and 1-5.

27. As is well known to mathematicians and statisticians, if the mathematical formula has enough variables in it, it is easy to fit a large number of curves to a set of data points. D. V. Gokahle, Ph.D., Department of Statistics, University of California, Riverside, personal communication.

28. U.S. Department of Labor, Occupational Safety and Health Administration, "Cancer Policy," *Federal Register* 45 (22 January 1980):5190.

29. For further discussion about some of these issues see *National Research Council, Risk Assessment in the Federal Government*, pp. 29–33.

30. Lee, D. Y., and A. C. Chang, "Evaluating Transport of Organic Chemicals in Soil Resulting from Underground Fuel Tank Leaks," in 46 Purdue Industrial Waste Conference Proceedings (Chelsea, Mich.: Lewis Publishers, 1992), pp. 131–140. If one assumes that ethylbenzene will be carried in moving water and that only 0.1 of the total solution is nonaqueous, then one will obtain predictions that 1500 times as much ethylbenzene will reach groundwater than if one assumes that the nonaqueous phase of the total liquid solution is moving and only 0.1 on the total solution is nonaqueous.

31. Ibid.

32. Personal communication from Andrew Chang. Also see Lee and Chang for some discussion of this point.

33. Robert Brown, mathematical statistician, Food and Drug Administration, personal communication and presentation at Pesticides and Other Toxics: Assessing Their Risks, University of California, Riverside, 17 June 1989.

34. Cothern, Coniglio, and Marcus, "Estimating Risks to Health."

35. A much more stable point would be the mean (average or expected value) estimate rather than the maximum likelihood estimate. Lauren Zeise, "Issues in State Risk Assessment: California Department of Health Services," *Proceedings: Pesticides and Other Substances: Assessing Their Risks,* ed. Janet L. White (Riverside: University of California, College of Natural and Agricultural Sciences, 1990) p. 142.

36. Dale Hattis, Massachusetts Institute of Technology, presentation, University of California, Riverside, 1988.

37. Lauren Zeise, senior toxicologist and acting chief, Office of Environmental Health Hazard Assessment Section, and Bill Pease, graduate student, University of California, Berkeley, Public Health School, personal communication.

38. Zeise, Wilson, and Crouch, "Dose–Response Relationships for Carcinogens," p. 266.

39. Allen, Crump, and Shipp, "Correlation Between Carcinogenic Potency of Chemicals," 531–34.

40. L. Zeise, W. Pease, and A. Kelter, "Risk Assessment for Carcinogens Under California's Proposition 65," address, Western Meetings of the American Association for the Advancement of Science, January 1989.

41. National Research Council, *Risk Assessment in the Federal Government*, pp. 17–50.

42. U.S. Cong., OTA, *Identifying and Regulating Carcinogens*, p. 25. Some of the particular inference gaps identified by the National Research Council report are summarized in Appendix A.

43. Ibid. For a general discussion of risk assessment policies and some of their history, see pp. 23–74 of the OTA study and Chapter 4 of this book.

44. Willard Van Orman Quine, *Word and Object* (Cambridge, Mass.: MIT Press, 1960), p. 78. He illustrates this point by reference to several examples. In mentalistic philosophy there is the familiar predicament of private worlds. In speculative neurology there is the circumstance that different neural hookups can account for identical verbal behavior. In language learning there is the multiplicity of individual histories capable of issuing identical verbal behavior (p.79).

45. I owe this way of putting the point to Larry Wright.

46. The U.S. Food and Drug Administration chose to use the Gaylor–Kodel high-dose to low-dose extrapolation procedures to ensure the decision procedure was independent of all models (with a claim to scientific accuracy) and easy to use. The Gaylor–Kodel procedure takes the last data point from an animal bioassay arrived at with confidence (or perhaps extrapolates from that with the 95% upper confidence linearized multistage model to the 25%

risk response) and draws a straight line through the origin with a ruler. Personal communication, David Gaylor, U.S. Food and Drug Administration.

47. See Howard Latin, "Good Science, Bad Regulation and Toxic Risk Assessment," *Yale Law Journal on Regulation* 5 (1988):89–142, for a summary of the many reasons for choosing various particular features of a risk assessment for benzene.

48. See Adam Finkel, "Is Risk Assessment Too Conservative: Revising the Revisionists," *Columbia Journal of Environmental Law* 14 (1989):427–67; J. C. Bailar, E. A. C. Crouch, R. Shaiklr, D. Spiegelman, "One-Hit Models of Carcinogenesis: Conservative or Not?" *Risk Analysis* 8 (1988):485–97; L. Zeise, "Issues in Risk Assessment," and Leslie Roberts, "Is Risk Assessment Conservative?" *Science,* 1989, p. 1553.

49. National Research Council, *Risk Assessment in the Federal Government,* p. 76.

50. Ibid.

51. Carl Cranor, "Some Public Policy Problems with the Science of Carcinogen Risk Assessment," *Philosophy of Science* 2 (1988):467–88. The EPA has apparently explicitly adopted this view as part of its concept of "risk characterization." The agency now acknowledges that characterizing a risk to which people may be exposed partly relies upon risk management or normative considerations. Person communication, Margaret Rostker, National Academy of Sciences, and the EPA Office of Pesticides and Toxic Substances.

52. One need only consider the recent debate about "cold" fusion to understand the importance of such caution.

53. David Dolinko and John Fischer pointed this out.

54. See Edward W. Lawless, *Technology and Social Shock* (New Brunswick, N.J.: Rutgers University Press, 1977), who describes a number of social and regulatory false negatives.

55. National Toxicology Program, Division of Toxicology Research and Testing, *Chemical Status Report* July 7, 1991.

56. U.S. Cong., OTA, *Identifying and Regulating Carcinogens,* p. 17. According to OTA, of "30 chemicals approved for testing in [1981 and 1982], 4 [had] reached the stage of chronic testing" by July 1987.

57. This modest estimate is taken from the National Research Council, *Toxicity Testing: Strategies to Determine Needs and Priorities* (Washington, D.C.: National Academy Press, 1984), p. 3.

58. Although the Toxic Substances Control Act is designed to screen substances before they go into commerce, the EPA has only 90 days to evaluate such substances or they automatically receive approval, unless EPA files for an extension of the evaluation period or requests additional testing.

59. Personal communication, risk assessors in the EPA's Office of Pesticides and Toxic Substances, July 1986. The discrepancy between the average calculated by OTA and the rate reported by EPA risk assessors is probably attributable to the fewer substances evaluated when the program was initiated. EPA officials indicated that 150 PMNs per month was a more typical number in 1987.

60. Interviews with federal risk assessors indicated that the average of 88/month might be somewhat low. They estimated that they might receive as many as 150/month. Personal communication, July 1986.

61. Zeise, "Issues in State Risk Assessment," pp. 139–41.

62. U.S. Cong., OTA, *Identifying and Regulating Carcinogens,* p. 20.

63. Fall 1990 information from the California Department of Health Services Office of Reproductive and Cancer Hazard Assessment Section. Available to members of the Proposition 65 Science Advisory Panel.

64. Estimate from Lauren Zeise, senior toxicologist and acting chief of the Reproductive and Cancer Hazard Assessment Section, Office of Health Hazard Assessment, California

Environmental Protection Agency, personal communication. A person year of effort is one person working for one year on the project or the equivalent, that is, two persons working for one-half year each.

65. Judith S. Mausner and Anita K. Bahn, *Epidemiology: An Introductory Text* (Philadelphia: W. B. Saunders, 1974), p. 320: Kenneth J. Rothman, *Modern Epidemiology* (Boston: Little, Brown, 1986), pp. 62–74.

66. Mausner and Bahn, *Epidemiology,* pp. 312–13; Rothman, *Modern Epidemiology,* pp. 62–74.

67. Rothman, *Modern Epidemiology,* p. 89.

68. On relative risk, Mausner and Bahn, *Epidemiology,* pp. 316, 318 (the estimation can be made only under certain assumptions, since the magnitude of the incidence rates cannot be derived.); on sample size, Rothman, *Modern Epidemiology,* pp. 79–82.

69. Rothman, *Modern Epidemiology,* pp. 64–68.

70. Mausner and Bahn, *Epidemiology,* p. 316. Unlike a retrospective cohort study.

71. Ibid., p. 93. These same problems plague retrospective cohort studies (discussed later).

72. D. Schottenfeld and J. F. Haas, "Carcinogens in the Workplace," *CA-Cancer Journal for Clinicians* 144 (1979):156–59.

73. Mausner and Bahn, *Epidemiology,* p. 320.

74. Ibid., pp. 323–24.

75. Ibid., pp. 324–25.

76. Schottenfeld and Haas "Carcinogens in the Workplace," pp. 156–59. See representative latency periods for various causes.

77. OSHA, *Federal Register* 45 (22 January 1980):5040.

78. Ibid.

79. Ibid.

80. Ibid., p. 5043.

81. Ibid., p. 5040.

82. Ibid., p. 5041.

83. Ibid., p. 5042.

84. Manolis Kogevinas and Paolo Boffetta in a letter to the *British Journal of Industrial Medicine* 48 (1991): 575–76, criticize a study by O. Wong, "A Cohort Mortality Study and a Case Central Study of Workers Potentially Exposed to Styrene in Reinforced Plastics and Composite Industry," *British Journal of Industrial Medicine* 47 (1990):753–62, for having too short a follow-up (seven years) even though he had large sample populations to study. They claimed that this defect "should caution against a premature negative evaluation of cancer risk in the reinforced plastics industry."

85. OSHA, *Federal Register* 45, p. 5044.

86. U.S. Congress, Office of Technology Assessment, *Cancer Risk: Assessing and Reducing Dangers in Our Society* (Boulder, Colo.: Westview Press, 1982), p. 137.

87. Ibid.

88. These points were suggested to me first by Debra Mayo and more recently by Shana Swan, chief, Reproductive Epidemiology, Office of Environmental and Occupational Epidemiology, California Department of Health Services.

For a literature discussion of these issues see A. M. Walker, "Reporting the Results of Epidemiologic Studies" (Different Views), *American Journal of Public Health* 76 (1986):556–58; J. L. Fleiss, "Significance Tests Have a Role in Epidemiologic Research: Reactions to A. M. Walker" (Different Views), *American Journal of Public Health* 76 (1986): 559–60; W. D. Thompson, "Statistical Criteria in the Interpretation of Epidemiologic Data" (Different Views), *American Journal of Public Health* 77 (1987): 191–94; C. Poole, "Beyond the Confidence Interval" (Different Views), *American Journal of Public*

Health 77 (1987): 195–99; S. N. Goodman and R. Royall, "Evidence and Scientific Research" (Commentary), *American Journal of Public Health* 78 (1988):1568–74.

89. E.A. Ades, "Evaluating Screening Tests and Screening Programs," *Archives of Disease in Childhood* 65 (1990):793.

90. S. D. Walter, "Determination of Significant Relevant Risks and Optimal Sampling Procedures in Prospective and Retrospective Comparative Studies of Various Sizes," *American Journal of Epidemiology* 105 (1977):387, 391 (Table 2). I do not discuss how the statistical variables are derived for a particular study from the raw data but only wish to show the conceptual relationships among them.

91. A. R. Feinstein, *Clinical Biostatistics* (St. Louis: C. V. Mosby, 1977):324–25.

92. The low value for α may also be a mathematical artifice explained historically. As Ronald Giere puts it, "The reason [for the practice of having a 95% confidence level to guard against false positives] has something to do with the purely historical fact that the first probability distribution that was studied extensively was the normal distribution." *Understanding Scientific Reasoning* (New York: Hall, Rinehart, and Winston), pp. 212–13. Two standard deviations on either side of the mean of a normal distribution encompasses 95% of the entire distribution. He adds, "95 percent is a comfortably high probability to take as the standard for a good inductive argument. Most scientists seem to think that science can get along with one mistake in 20, but not with too many more."

93. A concern to prevent confounding factors that might lead to false positives may be one of the main reasons for setting α so low. Such concerns seem to dominate recent discussions of problems in epidemiology. [Alvan R. Feinstein, "Scientific Standards in Epidemiologic Studies of the Menace of Daily Life," Science 242 (1988):1257–63.] For a sharp response to some of Feinstein's concerns see Sander Greenland, "Science Versus Advocacy: The Challenge of Dr. Feinstein," *Epidemiology* 2 (1991):64–72.

94. I have in mind diseases that are rare compared with common diseases such as measles or chicken pox. However, diseases such as leukemia are not as rare as angiosarcoma caused by vinyl chloride exposure or cervical cancer caused by exposure to diethylstilbestrol. The extreme rarity of the latter two diseases permitted their causal agent to be traced relatively easily.

95. Feinstein, *Clinical Biostatistics,* pp. 320–24.

96. Mausner and Bahn, *Epidemiology,* p. 322.

97. Figures are from Mausner and Bahn, quoting R. Doll and A.B. Hill, "Lung Cancer and Other Causes of Death in Relation to Smoking," *British Medical Journal* 2 (1956):1071, 1073.

98. Relative risk might also be misleading if a disease occurs only once in a million times among those exposed to substance x and otherwise almost never. Because of an almost vanishingly small denominator, the relative risk becomes very large, even approaching infinity.

99. The examples discussed in the remainder of this subsection are from C. Cranor, "Some Moral Issues in Risk Assessment," *Ethics* 101 (1990):123–43.

100. This is a complicated matter, for three times the normal risk of a certain kind of cancer may or may not merit further investigation. If the prevalence of disease in the general population is 1/100,000,000, the risk may not be worth study or regulation, unlike a disease with a prevalence of 1/100 or 1/1000.

101. Nearly all the numbers used in these examples are taken from C. Cranor "Epidemiology and Procedural Protections for Workplace Health in the Aftermath of the *Benzene Case,*" *Industrial Relations Law Journal* 5 (1983):372. Most of the figures for the examples are also summarized in the appendix of the same article. The numbers can also be calculated from equation (2) in Walter, "Determination of Significant Relative Risks," 387–88. Note

that even a large sample does not guarantee a sufficiently sensitive study if the follow-up period is too short compared with the latency period. See the discussion cited earlier between Wong and Kogevinas and Boffetta.

102. This difficulty with negative human epidemiological studies also applies to animal bioassays, which are essentially animal epidemiological studies. We should be similarly skeptical of "no-effect" results there.

103. The value of $\beta = .49$ is computed from equation (2) in Walter, "Determination of Significant Relative Risks," pp. 387–88. For technical details see also Cranor, "Epidemiology," p. 389. Some quantitative results from this equation are also summarized in Appendix B of Cranor, "Some Public Policy Problems," pp. 467–88.

104. A standard epidemiological text notes that "it is generally accepted in the medical literature that it is [only] safe to reject a null hypothesis when there is a less than five percent chance of being wrong (type I error)." A. Rimm, *Basic Biostatistics in Medicine and Epidemiology* (New York: Appleton-Century Croft, 1980) 201–2. Federal agencies that have considered the problem of the 95% rule generally, but not quite invariably, endorse its use. In 1979 the U.S. Interagency Regulatory Liaison Group in "Scientific Bases for Identification of Potential Carcinogens and Estimation of Risks," *Journal of the National Cancer Institute* 63(1) (1979):241–69, noted: "In the statistical evaluation of cancer incidence or mortality differences, there has been a strong tendency for particular confidence levels (e.g., 95%) and particular probability values (e.g., $p = 0.05$ or $p = 0.01$) to be used as standard points for finding of statistical significance . . . probability values fall along a continuum and should be so reported. The uniform use of a standard probability value is not suggested."

In 1986 the U.S. Interagency Staff Group on Carcinogens in "Chemical Carcinogens: A Review of the Science and Its Associated Principles," *Environmental Health Perspectives* 67 (1986) 201–82 noted: "The 0.05 p-value is a reasonable statistical bench mark with a history of use in the bioassay field. It would be imprudent to disregard it as an index of significance without good reason. It would be equally unwise to regard it as an absolute requirement or a sine qua non without considering the weight of biological evidence."

The different dates of these two reports and the subtle shift in 1986 to a greater emphasis on use of the 95% rule may signal a trend in the regulatory community toward emphasizing greater use of "good science" as a basis of regulation. This poses problems, as I argue below.

105. National Cancer Institute, Demographic Analysis Section, Division of Cancer Cause and Prevention, *Surveillance, Epidemiology and End Results: Incidence and Mortality Data 1973–1977* (monograph 57) (1981):662–63, Table 51.

106. From Appendix C.

107. The following table shows the number of subjects needed to calculate various values of relative risk for cohort and case-control studies of congenital heart disease, a relatively common disease whose prevalence is .008 (8/1000) in the general population. That table assumes $\alpha = .05$ and $\beta = .10$.

Relative Risk	Cohort Study	Case-Control Study
2	3,837	188
3	1,289	73
4	712	45
5	478	34
6	280	24
7	168	18

Taken from J. J. Schlesselman, "Sample Size Requirements in Cohort and Case-Control Studies of Disease," *American Journal of Epidemiology* 99(1974):381, 382–83 (Tables 2 and 3). Note that the requisite sample size would have to become larger as the disease becomes rarer. The numbers needed for a case-control study would, of course, be large for a disease with a background rate 1/100 of the disease modeled here.

108. Ralph I. Horowitz and Alvan R. Feinstein, "Methodologic Standards and Contradictory Results in Case-Control Research," *American Journal of Medicine* 66 (1979):556–67, Linda C. Mayes, Ralph I. Horowitz, and Alvan R. Feinstein, "A Collection of 56 Topics with Contradictory Results in Case-Control Research," *International Journal of Epidemiology* 17 (1988):680–85.

109. Talbot Page, "The Significance of P-Values" (forthcoming).

110. Our research suggests some amendments to Page's claims about some aspects of hypothesis testing and use of the 95% rule. The power of appropriate tests in his case is quite high, even though the test is not statistically significant by the 95% rule. The power for 17 control and treatment tumors when $\alpha = .0542$ is about .9468, while the power of the test for 16 treatment and control tumors is .9630 when $\alpha = .0857$. This point was established by the author and Bill Kemple working on the UCR Carcinogen Risk Assessment Project.

111. Results of representative animal bioassays are taken from research under the UCR Carcinogen Risk Assessment Project by Bill Kemple and Carl Cranor, and summarized in Appendix D in Cranor, "Some Public Policy Problems," pp. 467–88.

112. How serious a problem is presented by this asymmetry between traditional α and β values depends in part upon what moral theory you believe to be correct and the numbers of people exposed to the substance. It also is dependent upon the substances in question. We might think of the benefits and costs associated with chemicals tested in two different ways: Are the magnitudes of the benefits provided by the substances comparable to the magnitudes of the harms threatened or not? The magnitudes of the benefits are comparable to the magnitudes of the harms, when, for example, one of the benefits promised is the saving of lives (comparable to the harm threatened by carcinogens which take lives) or the prevention of death. When the magnitudes of the harms and benefits are comparable, then perhaps we should have a greater concern about the possibility of false positives, for we may lose substantial life-saving benefits if we falsely condemn a substance when an epidemiological study has a false positive outcome. Contrast this possibility with the results when we have a substance that does not promise life-saving or death-preventing benefits. For example, the Environmental Protection Agency recently issued an alert on Alar, a systemic chemical that is used on apples, peanuts, grapes, and a number of other fruits. Its primary benefit in apples is to prevent them from falling off trees too early, to keep them firm and ripe-looking longer, and to delay the onset of rotting. The benefits accrue almost totally in marketing and profitability, with no obvious beneficial health effects at all and seemingly no life-saving effects. The chief harm posed by Alar is that one of its by-products, UDMH, is a potent carcinogen. In an epidemiological study of Alar, using a small α value which ensures good science, even though the risks posed to our health by Alar may be substantial (because of its potency), such risks might not be detected because of the chances of false negatives. The general strategy for addressing these decision problems is discussed in Chapter 5.

113. This does not indicate the combined false positive rate of the two studies; that is a further and more complex issue.

114. Sander Greenland, "Invited Commentary: Science versus Public Health Action: Those Who Were Wrong Are Still Wrong," *American Journal of Public Health* 133(5) (1991):435–36.

115. H. J. Eysenck, "Were We Really Wrong?" *American Journal of Epidemiology* 133(5) (1991):429, 432.

116. This traditional approach was suggested to me by Michael Simpson and Christopher Hill of the Congressional Research Service of the Library of Congress, as well as several statisticians. More recently statisticians have avoided the use of set α and β values, Deborah Mayo, personal communication. See also the discussion in the articles noted in endnote 119 below.

117. See E. L. Lehman, *Testing Statistical Hypotheses* (New York: Wiley, 1959), pp. 60–63 for a brief discussion.

118. This alternative or a variation of it was suggested independently by Deborah Mayo and Ronald Giere and is developed in Deborah Mayo, ''Toward a More Objective Understanding of the Evidence of Carcinogenic Risk,'' *PSA 1988* 2 (1988):489–503.

119. See Walker, ''Reporting the Results of Epidemiologic Studies''; Fleiss, ''Significance Tests Have a Role in Epidemiologic Research''; Thompson, ''Statistical Criteria in the Interpretation of Epidemiologic Data''; Poole, ''Beyond the Confidence Interval''; Goodman and Royall, ''Evidence and Scientific Research.''

120. Goodman and Royall, ''Evidence and Scientific Research,'' p. 1568.

121. Poole, ''Beyond the Confidence Interval.''

122. Of course, there is also the possibility that allegiance to their employer may also affect this risk assessment and risk management judgments. See Francis M. Lynn, ''OSHA's Carcinogen Standard: Round One on Risk Assessment Models and Assumptions,'' in *The Social and Cultural Construction of Risk,* eds. B. B. Johnson and V. T. Covello (Bingham, Mass.: D. Reidel, 1987), pp. 345–58.

123. See Richard J. Pierce, Jr., ''Two Problems in Administrative Law: Political Polarity on the District of Columbia Circuit and Judicial Deterrence of Agency Rulemaking,'' *Duke Law Journal* (1988):300, who argues for this ''democratic feature'' of the agencies.

124. In *Gulf South Insulation v. Consumer Product Safety Commission,* 701 F.2d 1137 (5th Cir. 1983) the court clearly misunderstood a risk assessment done on urea–formaldehyde foam insulation and invalidated a regulation as a consequence. Whether this will be a problem in the future depends upon court approaches to reviewing agency decisions.

125. For a view similar to the one just articulated, consider the following statement by Sander Greenland, which I discovered just as this book went to press. ''My view is that epidemiologists can afford the luxury of unrestrained scientific skepticism when they function as pure scientists. Some epidemiologists, however, function as public health officials (such as the Surgeon General) in making public health decisions. In these public health roles, epidemiologists and the society they serve must recognize the decision to not act as one more potentially wrong and costly decision. This recognition must operate whether the decision concerns the recall of a food product, the regulation of a chemical exposure, the ban of a drug, or simply warning the public against a voluntary exposure such as smoking.'' Sander Greenland, ''Re: 'Those Who Were Wrong,' '' (letter) *American Journal of Epidemiology* 132(3): (1990).

126. One might be concerned that with relatively rare diseases (e.g., 8/10,000 or 8/100,000) the problems of epidemiological studies just indicated might be of lesser concern. Although this may be true for extremely rare diseases, e.g., 1/10,000,000, it is not so obvious for above rates typical of numerous cancers. (The incidence rate of leukemia in the general population is 8/100,000.)

127. See Cranor, ''Some Public Policy Problems,'' for some discussion of this point.

128. Lynn, ''OSHA's Carcinogen Standard,'' pp. 345–58.

129. The exact aim of a particular regulatory law depends upon the statute in question. Some statutes would ban a substance outright if it was found to cause cancer in human beings, e.g., the Delaney Clause of the Food, Drug and Cosmetic Act; others might ban a substance only if that was a ''reasonable'' way to prevent the risk of harm from a carcinogen, e.g., the

Federal Insecticide, Fungicide and Rodenticide Act; still others might ban a substance if such a ban was technologically and economically "feasible" way of protecting people, e.g., the Occupational Safety and Health Act.

130. Even here demanding standards of scientific evidence can be controversial, for satisfying them may delay the introduction of life-saving drugs while the appropriate evidence is provided. In such cases one must weigh the moral and social costs and benefits of possibly incurring false positives versus false negatives to see whether such stringent standards should be applied.

131. For example, regulatory agencies may nominate substances to the National Toxicology Program for testing simply because they appear to pose health risks. The testing that is done under this program is aimed at regulatory purposes, and while some additional basic research may be obtained from the results of such tests, that is not the primary aim. (Although the research done under this program typically experiments with animals, similar arguments would apply.) In fact, if the arguments of this book are correct, perhaps regulatory agencies should be very careful in applying the 95% rule for their purposes.

132. Typically, for successful prosecution of a case the tort law requires that the plaintiff support his position by a "preponderance of the evidence." If this could be quantified, it appears to mean that more than 50% of the evidence favor the plaintiff (some place the estimate as high as 65%). The burden-of-proof issues in torts are much more complex than this remark suggests and are discussed at length in Chapter 2.

133. This point was suggested by Michael Resnick. 134. In addition, there is a concern that if one conducts animal or human statistical tests for cancer at multiple sites in the target species, the overall chances of having a false positive rise dramatically. "The argument is that, if n independent tests are conducted at the α significance level, the probability of finding a significance somewhere (the false positive rate) is $[1-(1-\alpha)^n]$. For $n = 20$ and $p = .05$ for a single dose–sex comparison, this probability is about .64." U.S. Interagency Staff Group, "Chemical Carcinogens," p. 242. The same group reports that in fact the false negative rates were considerably lower in two different animal studies: 8% in one case, 21% in the other.

CHAPTER 2

1. References are too numerous to list, but see "Rethinking Tort and Environmental Liability Laws: Needs and Objectives of the Late 20th Century and Beyond," symposium, *Houston Law Review* 24 (1987):1–219, for a number of papers that discuss these issues and for references to other articles on the tort law.

2. Richard Abel, "A Critique of Torts," *UCLA Law Review* 37 (1990):785.

3. Peter Huber, "Safety and the Second Best: The Hazards of Public Risk management in the Courts," *Columbia Law Review* 85 (1985):277.

4. Richard J. Pierce, "Encouraging Safety: The Limit of the Tort Law and Government Regulation," *Vanderbilt Law Review* 33 (1980):1281, 1283.

5. Jeffrie G. Murphy and Jules L. Coleman, *The Philosophy of Law: An Introduction to Jurisprudence*, rev. ed. (Boulder, Colo.: Westview Press, 1989), pp. 109, 114–15.

6. Ibid., p. 115.

7. Contract law can provide some workplace protections from toxic substances, e.g., as part of union contract.

8. Murphy and Coleman, *The Philosophy of Law*, p. 115. Pierce, "Encouraging Safety," pp. 1284–88, indicates some other reasons.

9. Of course, either administrative rules guiding our behavior with considerable specificity, e.g., requiring the use of particular pollution control technologies, or the tort law aims

to modify and control our behavior. The difference is in the details of the guidance provided and in whether the guidance is provided in advance or merely consists in the threat of a lawsuit.

10. Lecture notes from Bruce Ackerman's environmental law course, Yale Law School, 1980.

11. W. P. Keeton, D. B. Dobbs, R. E. Keeton, and David G. Owen, *Prosser and Keeton on Torts,* 5th ed. (St. Paul, Minn.: West Publishing Co., 1984), pp. 534–83.

12. See William H. Rodgers, Jr., *Handbook on Environmental Law* (St. Paul, Minn.: West Publishing Co., 1977), Chapter 2, and Richard B. Stewart and James E. Krier, *Environmental Law and Policy* (New York: Bobbs-Merrill, 1978), Chapter 4, for some discussion of this.

13. Paul Brodeur, *Outrageous Misconduct* (New York: Pantheon, 1985); Irving Selikoff, Jacob Churg, and E. Cuyler Hammond, "Asbestos Exposure in Neoplasia," *Journal of the American Medical Association* 188 (1964):22.

14. Out of fairness on this issue, which agencies might possibly have had jurisdiction is not entirely clear, since so many were created in the early 1970s after the epidemiological evidence of asbestos-caused disease.

15. U.S. Congress, Office of Technology Assessment, *Identifying and Regulating Carcinogens* (Washington, D.C.: GPO, 1987), pp. 80–81.

16. Ibid.

17. Ibid.

18. "Asbestos Ban Cleared. EPA Official Criticizes OMB for Costly Delays," *Environmental Health Letter* 28(13) (1989):101.

19. David Rosenberg, "The Dusting of America, A Story of Asbestos—Carnage, Cover-Up and Litigation," *Harvard Law Review* 99 (1986):1693, 1695.

20. *Borel v. Fiberboard Paper Products Corp. et.al.,* 493 F.2d (1973) at 1076, 1083.

21. "The tort system emerged as *the* uniquely effective and indispensable means of exposing and defeating the asbestos conspiracy, providing compensation to victims and deterring future malfeasance. . . . Every other institutional safeguard [failed]: the asbestos companies, of course, but also the medical and legal professions, the unions, the insurance carriers, and all manner of regulatory and legislative bodies. To be sure the tort system is far from perfect; but if left to other devices the asbestos conspiracy would have been buried along with its victims." Rosenberg, "The Dusting of America," pp. 1693, 1695.

22. Ibid., p. 1704.

23. *Ferebee v. Chevron Chemical Co.,* 736 F.2d 1529 (D.C. Cir. 1984).

24. This case raises an important issue concerning the relation between regulatory actions and tort law remedies—whether such regulations should preempt state tort law. Some commentators find the court's decision troubling. Stephan D. Sugarman, "The Need to Reform Personal Injury Law Leaving Scientific Disputes to Scientists," *Science* 248 (1990):823–27.

25. *Ferebee v. Chevron.*

26. When I served as a congressional fellow during the 99th Congress (1985–86), there was considerable discussion among congressional staff that *one* of the reasons pesticide manufacturers were willing to agree to amendments to the Federal Insecticide, Fungicide and Rodenticide Act was that several tort suits decided against manufacturers had motivated them to support more protective pesticide legislation in order to reduce the likelihood of tort suits.

27. *Sindell v. Abbot Labs. et al.,* 26 Cal.3d 588 (1980) at 593.

28. Ibid., at 594.

29. Edward W. Lawless, *Technology and Social Shock* (New Brunswick, N.J.: Rutgers University Press, 1977), pp. 73–74.

30. Ibid., p. 75.

31. *Sindell*, at 594. Even though the FDA banned DES for pregnant women, in 1973 it approved a DES "morning after" anticonception pill as a prescription drug. The dosage in this pill was much greater than concentrations of DES residues in beef and poultry, which the FDA had banned in 1972. Lawless, *Technology*, pp. 79–80.

32. C. P. Gillette and J. E. Krier, "Risk, Courts and Agencies," *University of Pennsylvania Law Review* 138 (1990):1027, 1087, citing a study by G. Eads and P. Reuter, *Designing Safer Products: Corporate Responses to Product Liability Law and Regulation* (Santa Monica, Calif.: Rand Corporation, 1983).

33. The Bendectin cases are one object of study in the University of Houston's Environmental Liability project. Early cases resulted in plaintiff verdicts, whereas later cases produced defendant verdicts with judges taking a much more active role in deciding the cases via directed verdicts or judgments notwithstanding the verdict. Although most scientists who specialize in teratology do not seem convinced that Bendectin causes birth defects, some of the authors I interviewed are concerned about it and some think there may be an association between Bendectin use and birth defects.

34. J. L. Mills and D. Alexander, letter, *New England Journal of Medicine* 315 (1986):1234. One author (Watkins) of the original epidemiological study, which tentatively suggests an association between prenatal exposure to vaginal spermicides and birth defects, disavows the study; the remaining three authors continue to subscribe to it. Watkins indicates that there remains doubt about "limb reduction anomalies," although there appear to be "chromosomal alterations." A fifth author (Holmes) thinks the study should not have been published in its original form, because it has been misused. Richard N. Watkins, letter, *Journal of the American Medical Association* 256 (1986): 3095; Herschel Jick, Alexander M. Walker, and Kenneth J. Rathman, letter, Ibid., p. 3096; Lewis B. Holmes, letter, Ibid., p. 3096.

35. The evidentiary standards for subatomic physics may be much more demanding than those for geology, psychology, or social science statistical studies.

36. A common conceptual framework is sketched later in this chapter.

37. Barbara Underwood, "The Thumb on the Scales of Justice: Burdens of Persuasion in Criminal Cases," *Yale Law Journal* 86 (1977):1299, 1311. However, even though this suggests that slightly more than 50% of the evidence must favor the plaintiff's case, some judges would place the probabilities higher. See Underwood and also Fleming James and Geoffrey Hazard, *Civil Procedure*, 2d ed. (Boston: Little, Brown, 1977), p. 243.

38. This standard was *constitutionally* required under *In re Winship*, 397 U.S. 358 (1970).

39. Sugarman, "The Need to Reform Personal Injury Law," pp. 823–27, issues a more general criticism of the use of science in tort law. I do not consider his much more extensive criticisms, since they go well beyond the standards of proof required in the law.

40. Bert Black, "Evolving Legal Standards for the Admissibility of Scientific Evidence," *Science* 239 (1987):1511–12.

41. Ibid., 1511.

42. Bert Black, "A Unified Theory of Scientific Evidence," *Fordham Law Review* 56 (1988):595.

43. *In re "Agent Orange" Product Liability Litigation*, 611 F. Supp. 1223, 1231 (C.D. N.Y. 1985).

44. See *I Inside Litigation* 44 (September 1987).

45. *Johnson v. United States*, 597 F. Supp. 374 (D. Kan. 1984).

46. *The President's Council on Competitiveness* (Washington, D.C.: GPO, 1991), p. 8. (This report was issued as this book went to press, so the book does not address all the issues raised in the report in an integrated way.)

47. Ibid., p. 2.

48. *Ferebee,* at 1536 (emphasis added).

49. Black, "A Unified Theory," p. 611.

50. In the area of regulatory law and science, there is a substantial group of people engaged in toxic substances risk assessment who would require scientists and regulatory agencies proposing regulations to make sure that their regulations are based on the standards of good research science.

51. Testifying to the existence of causal connections even though their own disciplinary standards have not been met could lead to cognitive dissonance and possibly ostracism by professional colleagues. There is also a darker side to this attitude, which may influence some: substantial scientific research is funded by the industrial community, some members of which may be the objects of tort suits.

52. James and Hazard, *Civil Procedure,* p. 245.

53. The burden of production may exist for a particular issue in a case (e.g., for whether the plaintiff was contributorily negligent) and for the case in chief (e.g., whether defendant's act harmed the plaintiff).

54. In the discussion that follows I develop a stylized example of statistical evidence designed to illustrate some points about the burden of production and the introduction of expert testimony. Since this constructed example is just one piece of evidence, it may seem misleading. However the reader, now forewarned, should not be misled, and in many contexts important in toxic tort cases the example may not be too misleading. In later sections I also discuss more realistic situations.

55. Black, "A Unified Theory," p. 692.

56. James and Hazard, *Civil Procedure,* p. 241.

57. Ibid., pp. 251–53.

58. Underwood, "The Thumb on the Scales of Justice," p. 1311, quoting the findings of the Chicago Jury Project.

59. *Frye v. United States,* 293 F.2d 1013 (D.C. Cir. 1923).

60. Black, "Evolving Legal Standards," p. 1510.

61. Ibid.

62. See *Fed. R. Evid.,* 702, 703, and Cleary, E. W., Ball, V. C., Barnhart, R. C., Brown, K. S., Dix, G. E., Gelhorn, E., Meisenkalb, R., Roberts, E. F., & Strong, J. W., *McCormick on Evidence,* 2nd ed. (St. Paul: West Publishing Co., 1972), sec. 13, p. 29.

63. *Ferebee,* at 1534.

64. *Wells v. Ortho Pharmaceutical Corp.,* 788 F.2d 749 (1986).

65. Mills and Alexander, letter; Holmes, letter.

66. Watkins, letter; Jick, Walker, and Rathman, letter; Holmes, letter; see at note 34.

67. Black, "A Unified Theory."

68. *Johnson v. U.S.*

69. *Lynch v. Merrill-National Labs.,* 830 F.2d 1190 (1st Cir. 1987), and *Richardson by Richardson v. Richardson-Merrell, Inc.,* 649 F. Supp. 799 (D.D.C. 1986); *aff'd at* 857 F.2d 823 (D.C. Cir. 1989), *cert. denied* 110 S. Ct. 218 (1989).

70. Black, "A Unified Theory," p. 662.

71. *Bertrum v. Wenning,* 385 S.W.2d 803 (Mo. App. 1965).

72. Black, "A Unified Theory," p. 668.

73. Ibid.

74. Black, "Evolving Legal Standards," p. 1511.

75. Black, "Evolving Standards," Ibid. As I indicated above, however, it is not entirely clear how "well established" this claim was.

76. See the text at notes 65 and 66.

77. A judgment notwithstanding the verdict is a decision by a trial court or appellate court *judge,* upon a motion by either party to enter a judgment for the party who was entitled to it as a matter of law *upon the evidence,* where the verdict was in favor of the other party. James and Hazard, *Civil Procedure,* p. 340.

78. Black "Evolving Legal Standards," p. 1511.

79. Ibid. (emphasis added). Current rules about expert testimony appear to be one target of the President's Council on Competitiveness. The argument in that report appears to be that in part because of liberal expert witness rules, U.S. industries, which can be more easily sued and face expensive litigation, are less competitive in international markets. *Council on Competitiveness,* p. 4. However, of all the factors preventing U.S. industries from being competitive (e.g., insufficient long-term research and development investment, high-salaried labor markets, extremely cheap labor markets for competitiors), expert witness rules appear to be negligible.

80. But as noted earlier, in toxic tort suits it may be rare that there is such evidence.

81. I owe this characterization to Sanford E. Gaines, director, University of Houston Law Center Environmental Liability Project.

82. This issue is complicated, of course, because at some point legal rules can become so restrictive that they impose substantial opportunity costs on the community. Peter Huber, *Liability: The Legal Revolution and Its Consequences* (New York: Basic Books, 1988), argues this point.

83. Eads and Reuter, *Designing Safer Products,* p. 31.

84. Laurence K. Altman, "New Method of Analyzing Health Data Stirs Debate," *New York Times,* 21 May 1990, p. 31.

85. This is a complicated subject which we cannot pursue further in this chapter, although we return to it in Chapter 4.

86. A scenario such as this may have occurred in epidemiological investigations into the health effects of low-level radiation exposure. In 1975 Alice Stewart, among others, discovered carcinogenic effects 33% higher among atomic plant workers exposed to low levels of radiation than the general population. Yet this discovery drew "immediate and vitriolic criticism from other scientists." Keith Schneider, "Scientist Who Managed to 'Shock the World' on Atomic Workers' Health," *New York Times* (national edition), May 3, 1990, p. A20. However, preliminary indications now (13 years later) indicate she may have been correct. *Occupational Safety and Health Letter* 20, June 27, 1990, p. 105.

87. Many toxic exposures result from substances or pollutants that receive *no* administrative screening before entering commerce; thus there likely is even less evaluation of such substances.

88. The Food, Drug, and Cosmetic Act, 1906, 1938, amended 1958, 1960, 1962, 1968, § 409.

89. The U.S. Congress's General Accounting Office recently evaluated the "unreasonable risk" standard of the Toxic Substances Control Act. Although it cannot account for all the poor EPA performance under TSCA's § 6 (requiring EPA to regulate substances that cause unreasonable risks of harm to health or the environment) on this legislative standard, some of it is attributable to this.

90. As a general rule it seems agencies may rely on less demanding standards of evidence than those that obtain in science. Some statutes (e.g., the Occupational Safety and Health Act) explicitly acknowledge this possibility, while many other agencies may regulate substances as long as their actions are not "arbitrary and capricious." (We return to this in Chapter 4.)

91. *Ferebee,* at 1536.

92. Of course, the administrative agency that issued premarket approval could, depending upon its procedures, recall or reevaluate the substance; that is a contingent matter.

However, we are considering the tort law and it cannot perform its protective function on Black's proposal until demanding standards of evidence are met.

93. Eads and Reuter, *Designing Safer Products*.

94. Black, "A Unified Theory," p. 622.

95. Ibid., p. 625.

96. Ibid.

97. Ibid.

98. Ibid., p. 627.

99. Ibid., p. 692.

100. Black Ibid.

101. Ibid., p. 611. Currently, Rule 702 provides: "if scientific, technical, or other specialized knowledge will assist the trier of fact to understand the evidence or to determine a fact in issue, a witness qualified as an expert by knowledge, skill, experience, training, or education, may testify thereto in the form of an opinion or otherwise."

102. We should also note that his version of Rule 702 is quite at odds with at least one circuit court of appeals. The Third Circuit in *In re Paoli R.R. Yard PCB Litigation* overturning the exclusion of plaintiffs' scientific testimony held that "courts must err on the side of admission rather than exclusion." 916 F.2d 829, 857 (1990). I return to this point in conclusion.

103. Black, "A Unified Theory," pp. 625, 627.

104. Ibid., pp. 622–24.

105. Ibid., p. 625 (publication does not confer validity) and pp. 689–90 (arguing that although clinical ecology results have been published in good journals, this does not imply acceptance or scientific validity).

106. Black, p. 694.

107. He rightly acknowledges that publication is not sufficient evidence of valid reasoning.

108. David L. Hull, *Science as Process* (Chicago: University of Chicago Press, 1988), pp. 322–53.

109. Ibid., and personal communication.

110. This view is advanced by Hull. While some of his motivational attributions to scientists may not be fully adequate, his work is insightful.

111. Hull, *Science as Process,* p. 520.

112. Hull, personal communication, 27 July 1990.

113. A motion *in limine* is a "motion for an advance ruling on the admissibility of evidence . . . before the evidence is sought to be introduced." Cleary, et. al., *McCormick on Evidence,* p. 128.

114. Black, "A Unified Theory," p. 692. A motion *in limine* in one recent case on one expert witness lasted three days and probably duplicated much of the trial debate on scientific issues.

115. James and Hazard, *Civil Procedure,* pp. 231–32.

116. Cleary, et. al., *McCormick on Evidence,* p. 785.

117. Similar descriptions apply to the criminal law, only the probabilities are greater— the state must show "beyond a reasonable doubt" that its evidence has a much greater probability of being true than the defendant's evidence. Some have suggested that the state must establish its case by showing that 75–95% of the evidence favors its case. See Underwood, "The Thumb on the Scales of Justice."

118. I have benefited substantially from conversations with Kenneth Dickey about the results in this section.

119. This understanding of the maxim seems to agree with others' interpretations: David

Barnes, *Statistics as Proof* (Boston: Little, Brown, 1983), pp. 60–62, reports others using the same interpretation.

120. See some brief discussion of this point in Chapter 3 and Chapter 5.

121. Even this may concede too much importance to avoiding false positives, especially in toxic tort suits. If a false positive results a defendant will wrongly have to compensate, whereas if a false negative results an injured or killed plaintiff wrongly goes uncompensated. It is plausible that avoiding false negatives in torts is more important than avoiding false positives, but I do not rely on that point here. Furthermore, consider a common principle from torts: when courts have to make a decision under considerable uncertainty between a certainly negligent defendant who might not have actually *caused* plaintiff's harm and an innocent plaintiff, the defendant should bear the costs. Thus, according to that principle it appears more important to avoid false negatives than false positives, at least for negligently acting defendants. This is considered briefly in Chapters 3 and 5.

122. It seems as mistaken to have a "picture" of the world in which the "camera" has wrongly added objects to the scene as one in which the "camera" has wrongly omitted objects from the scene because it is too insensitive. Scientists have also expressed such views to me.

123. The argument of the preceding paragraphs permits a more general comparison between the error rates in science and the law, but it is not strictly needed for the concerns indicated here.

124. Kenneth Dickey suggested this simpler approach.

125. We can also arrive at the same conclusion [without utilizing (5)] by recognizing that the error rates embodied in the scientific study (which is the sole evidence in the hypothetical case) will determine the error rates of the tort case.

126. Herbert Needleman and David Billings, "Type II Fallacies in the Study of Childhood Exposure to Lead at Low Dose: A Critical and Quantitative Review" (forthcoming), in examining epidemiological studies of the effects of lead on the cognitive performance of children, found that false negative rates in these studies are sometimes as high as .82. That is, epidemiologists have eight chances out of ten of not detecting an association between lead exposure and reduced cognitive performance even if one exists! The false negative rates range from .48 to .82.

127. In addition, Talbot Page has shown that even though a statistical result is not significant at the .05 level, it may still provide good scientific evidence of a toxic effect. He shows that an animal bioassay that just misses being statistically significant at the .05 level according to a Fischer's exact test still provides evidence of toxic effects using Bayes's theorem. He also shows the two studies that just miss being significant at the .05 level on a Fischer's exact test, when combined (as these two properly could be), are statistically significant at the .05 level. Talbot Page, "The Significance of P-values" (unpublished).

128. *Sterling v. Velsicol Chemical Corp.*, 647 F. Supp. 303 at 499 (W.D. Tenn. 1986); *Sterling v. Velsicol Chemical Corp.*, 855 F.2d 1188 (6th Cir. 1988).

129. McCleary, et al., *McCormick on Evidence*, sec. 338, pp. 791–92.

130. Ibid., p. 791.

131. *Ferebee*.

132. Cleary et al., *McCormick on Evidence*, sec. 38, p. 791.

133. *Council on Competitiveness*, p. 4. Keeping the burden of production or the standard for expert testimony equal to or less demanding than the overall burden of persuasion in tort cases permits jurors to hear more evidence, in some cases less good evidence, but leave them to decide the issue.

134. Huber, "Safety and the Second Best," p. 277.

135. Furthermore, if the tort law, despite sending out "indistinct signals," affects companies' safety decisions more than the regulatory law, as Gillette and Krier as well as Eads and

Reuter (*Designing Safer Products*) note, this is a further reason for not unduly burdening the plaintiff. This is especially true if Gillette and Krier's thesis (just articulated) is correct.

136. Sanford E. Gaines brought this distinction to my attention.

137. A. Russell Localio, Lawthers, A. G., Brennan, T., Laird, N. M., Herbert, L. E., Peterson, L. M., Newhouse, J. P., Weiler, P. C., Hiatt, H. H. "Relation Between Malpractice Claims and Adverse Events Due to Negligence," *New England Journal of Medicine* 325 (1991):249.

138. Ibid., p. 250.

139. *Richardson by Richardson v. Richardson-Merrell, Inc.*, 857 F.2d 823, 832 (D.C. Cir. 1988), quoting *Ferebee*. Emphasis in original.

140. *Ferebee*.

141. 857 F.2d 923 (1988).

142. 916 F.2d 829 (1990).

143. 916 F.2d 829, 857 (1990). "Courts must err on the side of admission rather than exclusion."

144. *Sterling v. Velsicol Chemical Corp.* 855 F.2d 1188.

145. Ibid., p. 1217.

146. "New Jersey Supreme Court Applies Broader Test for Admitting Expert Testimony in Toxic Case," *Environmental Health Letter* 30 (27 August 1991):176.

147. Ibid.

148. *Rubanick v. Witco Chemical Corporation*, 576 A.2d 4 (N.J. Super. CT. A.D. 1990) at 15 (concurring opinion).

CHAPTER 3

1. Jack Nelson, "Critics Cry Foul on Helath, Safety," *Los Angeles Times*, 8 May 1983, p. 1.

2. *The Charlotte* (N.C.) *Observer*, 3–10 February 1980 (special Pulitzer Prize–winning issue on brown lung).

3. *Charlotte Observer*, Ibid.

4. *Charlotte Observer*, Ibid.

5. Sanford E. Gaines, "A Taxonomy of Toxic Torts," *Toxics Law Reporter*, 30 November 1988, p. 831.

6. Asbestos, vinyl chloride, and DES seem to be exceptions. Asbestos leaves the offending fibers behind; vinyl chloride and DES cause such rare diseases the cause is relatively easy to isolate.

7. Recent discoveries in genetics indicate the possibility that there may be genetic predispositions to certain diseases. See, for example, Johnny Scott, "Genes May Increase the Risk of Lung Cancer," *New York Times*, 1 August 1990, p. A11.

8. See Jeffrie G. Murphy and J. L. Coleman, *The Philosophy of Law: An Introduction to Jurisprudence* (Totowa, N.J.: Rowman and Allenheld, 1984), pp. 114–22, 167–207, for a discussion of the tort law and contrasts with the contract law and the criminal law.

9. See Herbert Packer, *The Limits of the Criminal Sanction* (Palo Alto, Calif.: Stanford University Press, 1968), pp. 24–25, and Steven Shavell, *Economic Analysis of Accident Law* (Cambridge, Mass.: Harvard University Press, 1987), pp. 278–86, for some characterizations of the differences among criminal, tort, and regulatory law. Some of these are discussed further below.

10. In general it is much easier to criticize preventive regulatory law than compensatory tort law. It is difficult to argue that if a person has been injured through the faulty conduct of

another (the typical complaint in torts), the victim should not receive compensation for harms suffered. On the contrary, it is relatively easier to argue that if only a few have been injured because of exposure to toxic substances, then it is costly and disutilitarian to impose preventive regulations merely to protect a small number of individuals. I believe that many who would oppose preventive regulations would not be as strongly opposed to tort law compensation for these reasons, provided all the elements of a tort action could be proved.

11. See W. D. Ross, "The Right and the Good," in *Readings in Ethical Theory,* 2d ed., W. Sellars and J. Hospers (New York: Appleton-Century-Crofts, 1970), p. 485, for a brief discussion of such a principle. One should note that these principles of liability for causation are not particularly radical, even though they are thought by some to impose special hardships on tort law plaintiffs in certain complex causal situations; see Richard Delgado, "Beyond Sindell: Relaxation of Cause-in-Fact Rules for Indeterminate Plaintiffs," *California Law Review* 70 (1982):881, 887.

12. I sometimes ignore detailed features of an area of law in order to clarify principles of liability for causation which are general enough to apply to wider contexts than those in which they are found.

13. Ronald Dworkin, *Law's Empire* (Cambridge, Mass.: The Belknap Press, 1986).

14. Keeton, W. P., D. B. Dobbs, R. E. Keeton, and D. G. Owen, *Prosser and Keeton on Torts*, 5th ed. (St. Paul, Minn.: West Publishing Co., 1984), p. 266.

15. Ibid.

16. Ibid., p. 267.

17. Ibid., p. 268.

18. *Anderson v. Minneapolis, St.P. & S.St.M. R.R. Co.*, 146 Minn. 430, 179 N.W. 45 (1920).

19. *Steinhauser v. Hertz Corp.*, 421 F.2d 1169 (2d Cir. 1970).

20. Ibid.

21. *Evans v. S.J. Groves and Sons Co.*, 315 F.2d 335 (2d Cir. 1963).

22. Ibid., at 347.

23. Ibid., at 348.

24. 445 P.2d 313 (Cal. Sup. Ct. 1968).

25. Ibid., at 319. The court further noted that "the decedent's employment need only be a contributing cause of his injury," and that the "employee's risk of contracting the disease by virtue of the employment must be materially greater than that of the general public. Thus, in Bethlehem we allowed an award to an employee who contracted a contagious disease, since he had shown that the disease was more common at his place of employment than among the public." And, finally, "although decedent's smoking may have been inadvisable, respondents offer no reason to believe that the likelihood of contracting lung cancer from the smoking was so great that the danger could not have been materially increased by exposure to the smoke produced by burning buildings."

There are two special features about this case. The city did not introduce evidence of its own to contradict the plaintiff's claim about the decedent's exposure to smoke on the job and the dangers therefrom; thus it is barely possible in a similar case in which such evidence was introduced that the decision might be different. However, the facts would seem to be against cities on such matters, since apparently inhaling smoke from burning buildings almost surely increases the risk of lung cancer. In addition, this is a workers' compensation case in which courts may be even more lenient toward plaintiffs than the tort law generally. Nevertheless, the court's reasoning strongly suggests the principle that where the employee's condition of work would substantially contribute to an injury like one he would bring on himself through his own actions, such as smoking, the employee or his heirs may be entitled to recover at least some workers' compensation damages for injuries suffered.

The applications are obvious. Had such a rule obtained in North Carolina many fewer workers would have gone uncompensated for byssinosis than *The Charlotte Observer* reports they did (note 2 above).

26. *Selewski v. Williams*, 72 Mich. App. 742, 250 N.W.2d 529 (1977).

27. Ibid.

28. *Elam v. Alcolac, Inc.*, 765 S.W.2d 42 (Mo. App. 1988) at 174.

29. Ibid., at 177.

30. Ibid.

31. *Restatement (Second) of Torts*, §§ 431, 432 comment A (1965).

32. *Elam v. Alcolac*, at 176.

33. Ibid., at 178.

34. Ibid., at 183.

35. Ibid., at 186, 187.

36. See C. P. Gillette and J. E. Krier, "Risk, Courts and Agencies," University of *Pennsylvania Law Review* 138 (1990):1027, 1055–58, who argue that procedural rules, including causation rules, favoring plaintiffs, merely help to balance out a bias *against plaintiffs* in obtaining *access* to tort remedies.

37. See the discussion in Chapter 2 and in Marc A. Franklin, *Cases and Materials on Tort Law and Aternatives* (Mineola, N.Y.: Foundation Press, 1979), pp. 2–18, for a discussion of the elements of a tort suit.

38. Barbara Underwood, "The Thumb on the Scales of Justice: Burdens of Persuasion in Criminal Cases," *Yale Law Journal* 86 (1977):1299. Of judges surveyed by the Chicago Law School Jury Project about half estimated the kind and amount of evidence needed to establish a case by a preponderance of evidence to be 55%; the remaining half estimated the percentages between 60 and 75 (p. 1311).

39. Keeton et al., *Prosser and Keeton on Torts*, p. 269. See also *Grinnell v. Pfizer Co.*, 274 Cal. App.2d 424, 435 (70 Cal. Rptr. 375, 1969).

40. See Estep, S.D. "Radiation Injuries and Statistics: The Need for a New Approach to Injury Litigation," *Michigan Law Review* 59 (1960):259, for a discussion of some of these issues.

41. See B. Black and D. E. Lillienfeld, "Epidemiologic Proof in Toxic Tort Litigation," *Fordham Law Review* 52 (1984):732, 761, for a discussion of how to calculate attributable risk.

42. Recently there have been several proposals to modify these proof requirements: Delgado, "Beyond Sindell" and David Rosenberg, "The Causal Connection in Mass Exposure Cases: A 'Public Law' Vision of the Tort System," *Harvard Law Review* 97 (1984):849. Relatedly, Duncan C. Thomas (University of Southern California, Department of Preventive Medicine) and some colleagues have proposed a procedure for identifying subpopulations of an exposed group that might have suffered leukemia as a consequence of fallout from atomic bomb testing in Utah in the 1950s. If these statistical methods prove fruitful, they may ease some of the proof burdens discussed above. See W. Stevens, J. E. Till, D. C. Thomas. J. L. Lyon, and R. A. Kerber, "A Historical Reconstruction and Epidemiologic Study of Leukemia in Utah Residents Exposed to Fallout"(forthcoming).

43. See Steven Shavell, "Liability for Harm Versus Regulation of Safety," *Journal of Legal Studies* 13 (1984):357.

44. Joel Feinberg, "Sua Culpa," in *Doing and Deserving*, ed. J. Feinberg (Princeton, N.J.: Princeton University Press, 1970), pp. 203, 204, 205.

45. See Guido Calabresi and Jon Hirschoff, "Toward a Test for Strict Liability in Torts," *Yale Law Journal* 81 (1972):1054–85, for such arguments.

46. Such actions would seem to be required by the legislative goals of the Occupational

Safety and Health Act of 1970, which requires the Secretary of Labor to set the standard that most adequately ensures, to the extent feasible, on the basis of the best available evidence, that no employee will suffer material impairment of health or functional capacity even if the employee has regular exposure to the hazard dealt with by this standard for the period of his working life. 29 U.S.C. § 655(b)(5) (1976).

47. And it could certainly rely upon the substantive standards of a statute such as the Occupational Safety and Health Act of 1970.

48. Shavell, *Economic Analysis of Accident Law,* p. 278.

49. Ibid.

50. Ibid., p. 283.

51. Ibid.

52. Ibid., p. 285.

53. See J. L. Coleman, "Strict Liability in Torts," *William and Mary Law Review* 18 (1976):259, for a discussion of this point.

54. See Black and Lillienfeld, "Epidemiologic Proof," p. 777, as well as K. L. Hall and E. K. Sibergeld, "Reappraising Epidemiology: A Response to Mr. Dore," *Harvard Environmental Law Review* 7 (1983):441, for a discussion of this point.

55. This assumes the kinds and amount of evidence remain the same for both areas of the law. However, regulatory law has tended to permit regulation based upon evidence that might be insufficient for the tort law preponderance of evidence rationale. See Black and Lillienfeld, "Epidemiological Proof," for a discussion of this point.

56. The U.S. Supreme Court, in *Industrial Union Dept., AFL-CIO v. American Petroleum Inst.,* 448 U.S. 607 (1980) at 653, seemed to hold that a showing such as this would be more than sufficient to establish regulations; whether it is also necessary is a matter of some controversy. See C. Cranor, "Epidemiology and Procedural Protections for Workplace Health in the Aftermath of the Benzene Case," *Industrial Relations Law Journal* 5 (1983): 372, for a discussion of some of the problems.

57. By imposing regulation upon firms and trying to prevent diseases before they occur this in effect forces firms to incorporate into the price of their products more nearly the full social costs, including the costs of disease and death, of producing them. At some point, of course, the costs of regulation can become too great to remain in the public interest.

58. Of course, workers might also be able to spread the costs of disease through insurance; this would not prevent disease but merely spread the compensation burdens after the fact. Thus, it does not seem a very desirable alternative.

59. Calabresi and Hirschoff, "Toward a Test for Strict Liability," p. 1055.

60. Or, to generalize to other environmental health statutes, the general public would also know little about the toxicity of substances.

61. In contract law, from which this argument is borrowed, the question is, Who should bear the risk of losing a lawsuit when factual uncertainty obtains with regard to a contract? The argument is that uncertainty is a cost, or at least resolving it is a cost, and an efficient legal system would impose the costs of resolving uncertainty on the party to the dispute who is better able to do so. Anthony Kronman, "Mistake and Impossibility: Arranging A Marriage Between Different Partners, *Northwestern University Law Review,* 74; pp. 166–201 (1978).

62. However, depending upon the burdens of proof that must be satisfied to regulate a substance, it might be easier or harder for an agency to regulate. If the burden of proof is on an agency under a postmarket statute, those exposed bear the costs of ignorance. If the burden of proof is on the industry, it bears the cost of uncertainty. California's citizen-passed initiative, Proposition 65, imposes the burden of proof on the risk creator to show that the public is

exposed to no significant risk of harm. This greatly facilitates regulation. These issues are considered in Chapter 4.

63. See U.S. Congress, Office of Technology Assessment, *Identifying and Regulating Carcinogens* (Washington, D.C.: GPO, 1987), pp. 199–200, for some discussion of this.

64. Were there space to develop this argument, we should distinguish between information gathering essential to doing competent epidemiological studies that a firm could do relatively easily and research into scientific knowledge about the nature of substances, their effects on human beings, and their method of transmission, all of which are more difficult to do. Both kinds of knowledge generation are important; the former, however, is much easier than the latter. And I think that in general firms should have the burden of providing both, but this cannot be developed further here.

65. Delgado, "Beyond Sindell," p. 894.

66. See C. D. Broad, "On the Function of False Hypotheses in Ethics," *International Journal of Ethics* (now known as *Ethics*) 26 (1916):377–97, for some discussion of this principle.

67. Ibid., pp. 388–89, for discussion of these aspects of the fairness principle.

68. One can see this point by recognizing that if an employee in a nuclear power plant were killed in an automobile accident, damage awards to his estate would be discounted by his reduced life expectancy because of exposure to atomic radiation.

69. *Smith v. Lockheed Propulsion Co.*, 247 Cal. App.2d 774 (56 Cal. Rptr. 128, 1967). (Defendant, who is engaged in the enterprise for profit, is in a position best able to administer the loss so that it will ultimately be borne by the public).

70. Franklin, *Tort Law and Alternatives*, p. 414; and Keeton et al., *Prosser and Keeton on Torts*, p. 557 (We might think of this as a principle for distributing risks in the community.)

71. This argument is not without its difficulties. For one thing, the rationale from ultrahazardous activities in torts is designed to protect people in the larger community from such risks of harm, and there is an explicit assumption-of-risk defense such that if a person voluntarily exposes himself to such ultrahazardous activities, e.g., a lion trainer, then he may not be able to recover. Keeton et al., *Prosser and Keeton on Torts*, p. 566. However, with the development of workers' compensation statutes, an employer's defense of assumption of risk was not permitted in allowing employees to recover for injuries suffered arising out of his employment. Ibid., p. 573. To further complicate the picture, the Tenth Circuit Court of Appeals upheld the trial court's decision in *Silkwood v. Kerr-McGee Corp.*, W.D. Okla. 1979, 485 F. Supp. 566, which permitted Karen Silkwood's estate to sue Kerr-McGee Corporation for injuries she probably suffered as a consequence of exposure to plutonium, even though she was an employee at the Kerr-McGee plant. Noted in Keeton et al., *Prosser and Keeton on Torts*, p. 558.

72. George Peters, of Peters and Peters Law Firm, Santa Monica, California, personal communication at the Third National Conference on Engineering Ethics (1983), at which an earlier version of portions of this work was presented.

73. See Shavell, *Economic Analysis of Accident Law*, pp. 277–78, for further discussion of this point.

74. Ibid., pp. 281–82.

75. Ibid., p. 282. This contrasts with the tort law, which imposes liability only after harm occurs (harm causing is not a necessary condition of criminal liability).

76. Ibid., p. 283.

77. See. C. Cranor, "Political Philosophy, Morality and the Law," *Yale Law Journal* 95 (1986):1066–86, for further discussion of this point.

78. See D. Rosenberg, "The Dusting of America, A Story of Asbestos—Carnage, Cover Up and Litigation," 1693–1706, esp. 1704–06, who argues that the tort law serves this "backup" function.

79. Although one may still want to place certain burdens of proof on industries to provide information about the toxicity of substances to keep accurate exposure and health records, etc.

80. In addition, it may well be that the costs of bringing tort suits for many people, including the costs of legal discovery and requisite proof of causation in environmental health suits, are not substantially less than the costs of administering the regulatory law.

81. Stephen D. Sugarman, "The Need to Reform Personal Injury Law Leaving Scientific Disputes to Scientists," *Science* 248 (1990):825–26.

82. Francis M. Lynn, "OSHA's Carcinogen Standard: Round One on Risk Assessment Models and Assumptions," in *The Social and Cultural Construction of Risk* ed. B. B. Johnson and V. T. Covello (Bingham, Mass.: D. Reidel, 1987), pp. 345–58.

The chemical Industry Institute for Toxicology, an industry-funded institute, was apparently begun in the late 1970s to address the scientific basis of OSHA's regulatory efforts. Interview with Karl Kronebusch, Office of Technology Assessment, March 1986.

83. Gillette and Krier, "Risk, Courts, and Agencies," p. 1086.

84. Ibid., p. 1087.

CHAPTER 4

1. This section draws upon aspects of Appendix I of U.S. Congress, Office of Technology Assessment, *Identifying and Regulating Carcinogens* (Washington, D.C.: GPO, 1987), pp. 199-220, which I wrote.

2. The Food, Drug and Cosmetic Act (FDCA), 1906, 1938, amended 1958, 1960, 1962, 1968. 21 U.S. C. §§ 301–94.

3. The Clean Air Act (CAA), 1970 amended 1974, 1977. 42 U.S.C. §§ 7401–7671q.

4. The Clean Water Act (CWA), formerly the Federal Water Control Act, 1972, amended 1977, 1978, 1987. 33 U.S.C. §§ 1251–1376.

5. The Occupational Safety Health Act (OSHA), 1970. 29 U.S.C. §§ 651–78.

6. The Safe Drinking Water Act (SDWA), 1976, amended 1986. 42 U.S.C. §§ 300f–300J-26.

7. The Resource Conservation and Recovery Act (RCRA), 1976. 42 U.S.C. §§ 6901–92. The Toxic Substances Control Act (TSCA), 1976. 15 U.S.C. §§ 2601–71.

8. The Comprehensive Environmental Response, Compensation and Liability Act (CERCLA or "Superfund"), 1980, amended 1986. 42 U.S.C. §§ 9601–75.

9. Richard A. Merrill, "Regulating Carcinogens in Food: A Legislator's Guide to the Food Safety Provisions of the Food, Drug and Cosmetic Act," *Michigan Law Review* 72 (1979):172–250.

10. 21 U.S.C. § 348(c)(3)(a).

11. There are three different versions of the Delaney Clause, the one just described for food additives, one enacted at the same time for drugs fed to animals, and one enacted in the Color Additives Amendments of 1960. Although they differ in detail, all prohibit animal and human carcinogens.

12. 21 U.S.C. § 376

13. 7 U.S.C. § 136j(a)(a)(a).

14. This means it "will perform its intended function without unreasonable adverse effects on the environment," and "when used in accordance with widespread and commonly

recognized practice it will not generally cause unreasonable adverse effects on the environment." 7 U.S.C. § 136a(c)(5). "Unreasonable adverse effects on the environment" means "any unreasonable risk to man or the environment, taking into account the economic, social and environmental costs and benefits of the use of any pesticide." 7 U.S.C. § 136(bb).

15. 15 U.S.C. §§ 2604(d)(1)(a), 2807(a)(2).

16. 15 U.S.C.§§ 2604(a)(2) (e), (f).

17. Safe Drinking Water and Enforcement Act of 1986; California Health and Safety Code, § 25249.5 (1989).

18. There are other possibilities, of course, but these seem less plausible; e.g., an agency might have the burden to show that a substance does not pose a risk before it enters commerce (a duty more naturally left to the manufacturer), or the manufacturer might have the burden to show that a substance poses a risk of harm (highly implausible and one a manufacturer would not undertake with enthusiasm).

19. 5 U.S.C. § 553(b).

20. 5 U.S.C. § 553(c).

21. Jerry L. Mashaw and Richard A. Merrill, *Administrative Law: The American Public Law System,* 2d ed. (St. Paul, Minn: West Publishing Co., 1985), pp. 416–23.

22. Ibid., p. 259.

23. 42 U.S.C. § 7412(a)(2); 33 U.S.C.§ 1317(a)(4).

24. 42 U.S.C. §§ 300g–(b)(1)(b).

25. Merrill, "Regulating Carcinogens in Food," p. 172.

26. U.S. Congress, Office of Technology Assessment, *Assessment of Technologies for Determining Cancer Risks from the Environment* (Washington, D.C.: GPO 1981), p. 180.

27. Ibid.

28. For example, the Food and Drug Administration has no need for risk assessment under the Delaney Clause, for even if the smallest amount of a direct food additive causes cancer in laboratory animals if it is found "unsafe" and banned as an additive.

29. John P. Dwyer, "The Limits of Environmental Risk Assessment" *Journal of Energy Engineering* 116(3) (1990):231–2.

30. U.S. Cong., OTA, *Identifying and Regulating Carcinogens,* p. 30.

31. Ibid., p. 31.

32. Ibid., pp. 33–34.

33. Ibid., p. 34.

34. Ibid., p. 35.

35. 29 U.S.C. § 655(b)(5).

36. *Industrial Union Dept., AFL-CIO v. American Petroleum Inst.,* 448 U.S. 607 (1980).

37. OTA, *Identifying and Regulating Carcinogens,* p. 35.

38. Dwyer, "Limits."

39. Ibid.

40. Ibid.

41. Richard C. Schwing, "SRA: An Antidote to Symbols of Insanity," *RISK Newsletter,* January 1989, p. 2.

42. *Industrial Union, AFL-CIO v. American Petroleum Inst.,* at 642, 653.

43. See C. Cranor, "Epidemiology and Procedural Protections for Workplace Health in the Aftermath of the *Benzene* Case," *Industrial Relations Law Journal* 5 (1983):373, for some references.

44. OTA, *Identifying and Regulating Carcinogens,* p. 735.

45. Elizabeth Grossman, "Issues in Federal Carcinogen Risk Assessment: Occupational Safety and Health Administration," in *Proceedings: Pesticide Use in a Technological Society*

and Pesticides and Other Toxics: Assessing Their Risks, ed. Janet White (Riverside: University of California, College of Natural and Agricultural Sciences, 1990), p. 120.

46. Ibid.

47. John P. Dwyer, "Overregulation," *Ecology Law Quarterly* 15 (1988):729 characterizing John M. Mendelhoff's position from his book *The Dilemma of Toxic Substance Regulation: How Overregulation Causes Underregulation at OSHA* (Cambridge, Mass.: MIT Press, 1988).

48. Dwyer holds this view.

49. *Gulf South Insulation v. Consumer Product Safety Commision,* 701 F.2d 1137 (5th Cir. 1983), at 1145 (The in-home studies of people focused only on complaint residences, not on average or randomly selected residences, and epidemiological studies were insufficient to sustain the agency's conclusion.), 1146.

50. Ibid., at 1146.

51. Ibid.

52. Typical animal studies consider 50–100 controls (male and female) and 50–100 animals at two or three different exposure levels, thus typically using 150–400 animals per study.

53. *Gulf South Insulation v. Consumer Product Safety Commission,* at 1146.

54. Before the Supreme Court's decision concerning the same OSHA regulations in *Industrial Union Department, AFL-CIO v. American Petroleum Institute.*

55. Rosemary O'Leary, "The Impact of Federal Court Decisions on the Policies and Administration of the U.S. Environmental Protection Agency," *Administrative Law Review* 41 (1989):554.

56. Ibid., p. 561.

57. Ibid., pp. 564–66.

58. Ibid., p. 567.

59. Ibid., pp. 567–68.

60. Ibid., p. 569.

61. Ibid., pp. 567–68.

62. Ibid., p. 736.

63. R. Reagan, Federal Regulation, Executive Order No. 12,291, *Federal Register* 46 (17 February, 1981):13,193 and R. Reagan, Regulatory Planning Process, Executive Order 12,498, *Federal Register* 50 (4 January, 1985):1036; Dwyer, "Overregulation," p. 729; U.S. Cong., OTA, *Identifying and Regulating,* pp. 139–42.

64. Dwyer, "Overregulation," p. 729, and Grossman, "Issues," p. 122.

65. Grossman, "Issues," p. 122.

66. See U.S. Cong., OTA, *Identifying and Regulating,* pp. 23–24, for a discussion of this issue.

67. Personal communication, risk assessors at the U.S. Consumer Product Safety Commission, July 1986.

68. Grossman, "Issues," pp. 121–22.

69. However, in several recent regulatory actions OSHA has in fact used typical models for potency assessments, e.g., the linearized multistage high-dose to low-dose extrapolation model and body weight interspecies scaling factors, even though it may not be committed to these as a matter of policy.

70. Carl Cranor, Lauren Zeise, William Pease, Laura Weiss, Mark Hennig and Sara Hoover, "Improving the Regulation of Carcinogens: A Proposal to Expedite the Estimation of Cancer Potency" (forthcoming, *Risk Analysis).*

71. David Freedman and Hans Zeisel, "From Mouse to Man: The Quantitative Assessment of Cancer Risks," *Statistical Science* 3 (1988):3–56.

72. Ibid., p. 24.

73. Ibid.

74. Elizabeth L. Anderson, "Scientific Developments in Risk Assessment: Legal Implications," *Columbia Journal of Environmental Law* 14 (1989):411–25.

75. Ibid., pp. 415–17, 419, 421, 422.

76. Cranor, Zeise, Pease, Weiss, Hennig and Hoover, "Improving the Regulation of Carcinogens."

77. Ibid.

78. Ibid. As was just discussed, some of this documentation may be a response to or a preemptive capitulation to court scrutiny.

79. Ibid.

80. The practice just described has been typical of the California Department of Health Services, an agency that has evaluated 78 substances in 4 years, a record far superior to any federal agency.

81. Cranor, Zeise, Pease, Weiss, Hennig and Hoover, "Improving the Regulation of Carcinogens."

82. "Real" risks are those based upon retrospective accident and death data, while "theoretical" risks, according to critics, are those based upon projections of harm relying upon models.

83. On the Freedman and Zeisel extreme reading of the scientific support for risk assessment, there could be *no* risk assessments, for the scientific information is too incomplete.

84. Black's view is not quite as extreme as I imagine this one to be because he requires only that the scientific view published in the literature. Approximate results or tentative views may be published in the journals. The complete and accurate science view presupposes both completeness (no inference gaps) and accuracy (no approximations).

85. U.S. Cong., OTA, *Identifying and Regulating,* pp. 199–200.

86. In *some* respects the Delaney Clause of the Food, Drug, and Cosmetic Act authorizes the regulation of carcinogens on policy grounds. If a substance is found in appropriate tests to cause cancer in humans or experimental animals, it may be banned from the market. There is no need in such cases for a detailed risk assessment for the extent of risk is not a consideration. The FDA tried to modify the Delaney Clause to permit constituents of substances that posed only a *de minimis* risk to the public to enter (or stay) in commerce. The Court of Appeals for the District of Columbia did not permit this interpretation of the clause. *Public Citizen v. Young,* 831 F.2d 1108 (D.C. Cir. 1987).

87. Myra Karstadt, "Quantitative Risk Assessment: Qualms and Questions," paper presented at Grappling with Risk Assessment: On the Frontier of Science and the Law, Lake Tahoe, California, October 6–8, 1986.

88. Robert L. Seilkin, "Benzene and Leukemia: Opposed to the Shortcomings in the Current Regulatory Process" and Robert C. Barnard, "An Industry View of Risk Assessment," two papers presented at Grappling with Risk Assessment: On the Frontier of Science and Law, Lake Tahoe, California, October 6–8, 1986; Anderson, "Scientific Developments," pp. 411–25.

89. Some descriptions of the FDA policy, while acknowledging the agency must err on the side of public safety, appear to come close to the science intensive view. Mary Frances Lowe, "Risk Assessment and the Credibility of Federal Regulatory Policy: An FDA Perspective," *Regulatory Toxicology and Pharmocology* 9 (1989):131–40, esp. p. 136.

90. Seilkin, "Benzene and Leukemia"; Barnard, "An Industry View."

91. Ibid.

92. National Research Council, *Pharmacokinetics in Risk Assessment: Drinking Water and Health* (Washington, D.C.: National Academy Press, 1987), p. xi.

93. For some discussion of these issues see Adam M. Finkel, "Is Risk Assessment Too Conservative: Revising the Revisionists," *Columbia Journal of Environmental Law* 14 (1989):427–67 and Lauren Zeise, "Issues in State Risk Assessments: California Department of Health Services," in *Proceedings: Pesticides and Other Toxics: Assessing Their Risks,* ed. Janet L. White (Riverside: University of California, College of Natural and Agricultural Sciences, 1990), pp. 135–44.

94. National Research Council, *Pharmacokinetics* p. xi.

95. Cranor, Zeise, Pease, Weiss, Hennig and Hoover, "Improving the Regulation of Carcingens."

96. See National Research Council, *Pharmacokinetics* for a collection of articles on the subject.

97. Catherine Elgin, "The Epistemic Efficacy of Stupidity," *Synthese* 72 (1988):304.

98. "The problem is this: the more distinctions a system of categories admits, the less difference there is between categories. As we refine our conceptual schemes, we increase our chances for error." Ibid., p. 305.

99. National Research Council, *Pharmacokinetics* p. xi (emphasis added).

100. Raymond Neutra, chief of Epidemiology, California Department of Health Services, personal communication.

101. National Research Council, *Pharmacokinetics,* p. 474.

102. Craig Byus, University of California, Riverside, Department of Biochemistry–Biomedical Sciences, and member of the California Air Resources Board Science Advisory Panel, personal communication. Also, C. Cranor, C. Byus, A. Chang, D. V. Gokale, and L. Olson, Final Report to the University of California Toxic Substances Research and Teaching Program on the University of California, Riverside, Carcinogen Risk Assessment Project, August 1989.

103. See Lars J. Olson, "The Search for a Safe Environment: The Economics of Screening and Regulating Environmental Hazards," *Journal of Environmental Economics and Management* 19 (1990):1–18, for a theory to help decide such matters.

104. National Research Council, *Risk Assessment in the Federal Government: Managing the Process* (Washington, D.C.: GPO, 1983), p. 77. "Guidelines very different from the kinds described could be designed to be devoid of risk assessment policy choices. They would state the scientifically plausible inference options for each risk assessment component without attempting to select or even suggest a preferred inference option. However, a risk assessment based on such guidelines (containing all the plausible options for perhaps 40 components) could result in such a wide range of risk estimates that the analysis would not be useful to a regulator or to the public. Furthermore, regulators could reach conclusions based on the ad hoc exercise of risk assessment policy decisions" (p. 77).

105. 701 F.2d 1137 (5th Cir. 1983).

106. Suppose an agency must show that it is more probable than not that there is a significant health risk from exposure to a substance before it is justified in issuing new lower standards for that substance, the standard OSHA must satisfy since the decision in *Industrial Union Department, AFL- CIO American Petroleum Institute*. If such evidence is based on the results of animal bioassays extrapolated to human beings, then it may be technically impossible to satisfy this standard when large uncertainty ranges are present. With such uncertainties, how can an agency justifiably claim that its conclusion is "more probably than not" correct? Should regulation or tort law recovery not be permitted? The answer depends on one's philosophical views: Who should bear the cost of such uncertainties, the possible victims or

the regulated industry, its shareholders, and the consumers of its products? This example suggests that a conflict exists between the fundamental evidentiary standards of science and of the law, and that there may be a need to assign the burden of uncertainty to parties even at the level of the risk assessment. See Fleming James and Geoffrey Hazard, *Civil Procedure*, 2d ed. (Boston: Little, Brown, 1977), p. 241; Carl Cranor, "Some Moral Issues in Risk Assessment," *Ethics* 10 (1990):123–43.

107. The first and second reasons for the accurate science view indicated in the text are implicitly normative. Biological correctness and scientific accuracy, while not appearing to be normative considerations, in the regulatory context are normative because of the substantial effects on outcomes that their adoption would suggest.

108. National Research Council, *Pharmacokinetics* p. xi.

109. Ibid.

110. Arthur Furst suggested some very restrictive criteria for *scientifically* identifying *human* carcinogens. These include clear results in good animal studies of several species, epidemiological studies that provide good evidence of *causation,* and confirmation in several short-term tests. Arthur Furst, "Yes, But Is It a Human Carcinogen?" *Journal of the American College of Toxicology* 9 (1990):1–18. Yet by his account there is sufficient evidence for identifying only about 20 human carcinogens on (the best) scientific grounds. Identifying such "real risks" would be quite slow and would lag many years behind animal studies that identify probable human carcinogens, possibly leaving many people exposed to harmful substances. Arthur Furst, Fourth Annual Research Symposium, University of California Toxic Substances Research and Teaching Program, Santa Barbara, California, November 3, 1990.

111. California Health and Safety Code, §§ 25249.5 et seq.

112. These specifications of major and minor underregulation and overregulation as well as "accurate" results concede a lot to opponents of the views for which I argue. Given the many uncertainties in risk assessment, one might tolerate larger ranges of "errors."

113. Lester B. Lave and Gilbert S. Omenn, "Cost Effectiveness of Short-Term tests for Carcinogenicity," *Nature* 324 (1986):29–34; Lester B. Lave, Gilbert S. Omenn, Kathleen D. Heffernan, and Glen Drouoff, "A Model for Selecting Short-Term Tests for Carcinogenicity," *Journal of the American College of Toxicology* 2 (1983):126, 128; Lester B. Lave and Gilbert S. Omenn, "Screening Toxic Chemicals: How Accurate Must Tests Be?" *Journal of the American College of Toxicology* 7 (1988):567, 571; Lester B. Lave, Fanny K. Ennever, Herbert S. Rosenkranz and Gilbert S. Omenn, "Information Value of the Rodent Bioassay," *Nature* 336 (1988):631-33.

114. A. Fisher, L. Chesnut, D. Violette, *Journal of Policy Analysis Management* 8:88 (1989) and J. V. Hall, A. M. Winer, M. T. Kleinman, F. T. Lurman, V. Bonojur, and S. D. Colome, "Valuing the Health Benefits of Clean Air," *Science* 255:812–17 (1992). See Hall, et al. for further references to the value of life studies.

115. Hall, Ibid., and Shulamit Kahn, "Economic Estimates of the Value of Human Life," in *Ethics and Risk Asssessment in Engineering,* ed. Albert Flores (New York: University Press of America, 1989), pp. 57–72, reprinted from *IEEE Technology and Society Magazine* 5(2) (1986). The argument need not rest on the dollar value of mistakes, but could be understood as a loss of social utility, a broader concept.

116. Cranor, *et al.,* "Improving the Regulation of Carcinogens."

117. There is no logical connection between one's commitment to the efficient use of resources, in a nontechnical sense, and utilitarianism, but many who hold one view also hold the other. See Kristin Shrader-Frechette, *Science Policy, Ethics and Economic Methodology* (Boston: D. Reidel, 1985), Chapter 1.

118. See Amartya Sen and Bernard Williams, eds., *Utilitarianism and Beyond* (New York, Cambridge University Press, 1982), for some discussion of these issues

119. The theory of a single executive controlling the different agencies is not well instantiated in practice.

120. Bruce Ames uses a similar rationale to claim we should have used more resources to discourage self-inflicted but voluntary harms caused by drinking alcoholic beverages and smoking, both of which cause cancer. Although this has much to recommend it, he appears to blur the distinction between risks voluntarily assumed (with substantial knowledge of their effects) and those that are involuntary and often invisible. These distinctions are clearly relevant to an adequate public policy to reduce the risks of harm in our lives. See Bruce Ames, "Ranking Possible Carcinogenic Hazards," *Science* 236 (1989):221. For an alternative view see Carl Cranor, "Comparative Risk Judgments," *Protecting Drinking Water at the Source* (Davis: University of California Water Resources Center, 1991), pp. 151–64 for an alternative view.

121. Issues of justice can, of course, also arise from people suffering from a naturally caused disease, if the victims cannot afford treatment, or if they were deliberately excluded from health care by individual discrimination or governmental politics.

122. Paul C. Slovic, "Perception of Risk," *Science* 236 (1987):280–85.

123. Cranor, "Comparative Risk Judgments."

124. We can easily imagine cases in which we can save ten people by killing one person (who is without family and friends and who is near the end of his life) and harvesting his organs to use in life-saving transplants for the ten. Similarly, it is controversial to kill one person or let one person die in order to save ten others from death by identical means. See Samuel Scheffler, *The Rejection of Consequentialism: A Philosophical Investigation of the Considerations Underlying Rival Moral Conceptions* (New York: Oxford University Press, 1982), for some discussion of these issues.

125. See John Taurek, "Should the Numbers Count?" *Philosophy and Public Affairs* (1978); John T. Sanders, "The Numbers Should Sometimes Count," *Philosophy and Public Affairs* 17 (1988):3–14.

126. A recent General Accounting Office study on another topic quotes the Chemical Manufacturers Association as saying that the "unreasonable risk standard" of Section 6 of the Toxic Substances Control Act has not been as effective as it might have been, because the EPA "has been overly cautious about developing 'perfect' rules and as a result has devoted a significant portion of its limited resources to one rule." U.S. General Accounting Office, *Toxic Substances: Effectiveness of Unreasonable Risk Standards Unclear* (GAO-RCED-90–189).

127. Even I am not considering literally *all* social consequences. For example, in Chapter 2 and here I do not consider the effects of my proposals on the costs of insurance for firms. Nor do I consider other more global strategies for regulating toxic substances which might have fewer bad consequences. The inquiry must stop somewhere, and these insurability and other concerns—although not irrelevant—are beyond this essay.

128. Colin Park, of Dow Chemical Company, at a recent conference "OMB v. the Agencies," "emphasized that risk assessment is a regulatory tool that, while in need of improvement, is not broken." Harvard University School of Public Health, Center for Risk Analysis, *OMB v. the Agencies: The Future of Cancer Risk Assessment* (Cambridge, Mass., June 1991). In his remarks Dr. Park seems to endorse the use of policies in risk assessment and even appears to accept the "fundamental science policy assumptions used by agencies" (p. 13).

129. National Research Council, *Risk Assessment in the Federal Government,* p. 76.

130. Howard Latin, "Good Science, Bad Regulation and Toxic Risk Assessment," *Yale*

Law Journal on Regulation 5 (1988):89–142. Social policy considerations must play as prominent a role in the choice of risk estimates as in the ultimate determination of which predicted risks should be deemed unacceptable.

131. A large number of commentators (too numerous to list) across several fields support the view that risk assessment has several normative components.

132. James M. Humber and Robert F. Almeder, eds., *Quantitative Risk Assessment: Biomedical Ethics Reviews* (Clifton, N.J.: Humana Press, 1988); Allen, B. C., Crump, K. S., and Skipp, A. M., "Correlations Between Carcinogenic Potency of Chemicals in Animals & Humans," *Risk Analysis,* pp. 531–44 (Dec. 1988).

133. Latin, "Good Science," p. 91, and Cranor, "Some Public Policy Problems."

134. Latin, "Good Science: 91; Carl Cranor, "Some Public Policy Problems with the Science of Carcinogen Risk Assessment," *Philosophy of Science Association* 2 (1988):467–88; Cranor, "Some Moral Problems with Risk Assessment"; and Carl Cranor and Kurt Nutting, "Scientific and Legal Standards of Statistical Evidence in Toxic Tort and Discrimination Suits," *Law and Philosophy* 9 (1990):115–156.

135. Latin, "Good Science," p. 90.

136. Ibid., pp. 93–94 (emphasis added).

137. William Farland, "Issues in Federal Carcinogen Risk Assessment: Environmental Protection Agency," in *Proceedings: Pesticides and Other Toxics: Assessing Their Risks,* ed. Janet L. White (Riverside: University of California, College of Natural and Agricultural Sciences, 1990), pp. 127–36.

138. Latin, "Good Science"; Cranor, "Epidemiology and Procedural Protections"; "Some Problems with the Science of Risk Assessment"; "Some Moral Issues in Risk Assessment."

139. Latin, "Good Science," p. 137, and Cranor, "Epidemiology and Procedural Protections."

140. Latin, "Good Science," p. 138.

141. Ibid., pp. 138–42; Talbot Page, "A Generic View of Toxic Chemicals and Similar Risks," *Ecology Law Quarterly* 7 (1978):207.

142. Latin, "Good Science"; Dale Hattis and John A. Smith, Jr., "What's Wrong with Risk Assessment?" in *Quantitative Risk Assessment: Biomedical Ethics Reviews,* ed. James M. Humber and Robert F. Almeder (Clifton, N.J.: Humana Press, 1986), pp. 57–106.

143. Latin, "Good Science," p. 125.

144. Grossman "Issues," pp. 121–22.

145. Latin, "The Feasibility of Occupational Health Standards."

146. Hatlis and Smith, "What's Wrong with Risk Assessment?" p. 95.

147. John Broome, "Utilitarianism and Expected Utility," *Journal of Philosophy* 84 (1987):405–22, for some discussion of this point.

148. Cranor, *et al.,* "Improving the Regulation of Carcinogens."

149. Lauren Zeise, senior toxicologist and acting chief, Reproductive and Cancer Hazard Assessment Section, Office of Health Hazard Assessment, California Environmental Protection Agency.

150. Agencies should also look for other ways to expedite risk assessments, such as reducing the periods in which interested parties have to comment on regulations. Craig Byus, member of California's Air Resources Board Science Advisory Panel, indicates that this committee is adopting such changes.

151. Early work on this appeared in Cranor, "Epidemiology and Procedural Protections."

152. 448 U.S. at 690, 716.

153. Ibid., at 706.

154. Ibid., at 707.

155. Ibid., at 693.

156. Ibid., at 707–8 (emphasis added).

157. Ibid., at 706.

158. I do not consider the environmental aspects of risk assessment, which can also be quite time-consuming.

159. Much of the discussion in this section is taken from Cranor, *et al.,* "Improving the Regulation of Carcinogens."

160. R. Peto, M. Pike, L. Bernstein, L. Gold, and B. Ames, "The TD_{50}: A Proposed General Convention for the Numerical Description of the Carcinogenic Potency of Chemicals in Chronic-Exposure Animal Experiments," *Environmental Health Perspectives* 58 (1984):1–8.

161. L. Gold, C. B. Sawyer, R. Magaw, G. Backman, M. de Veciana, R. Levinson, N. Hooper, W. Havender, L. Bernstein, R. Peto, M. Pike, and B. Ames, "A Carcinogenic Potency Database of the Standardized Result of Animal Bioassays," *Environmental Health Perspectives* 58 (1984):9–319.

162. Ibid.; L. Gold, M. de Veciana, G. Backman, R. Magaw, P. Lopipero, M. Smith, M. Blumental, R. Levinson, L. Bernstein, and B. Ames, "Chronological Supplement to the Carcinogenic Potency Database: Standardized Results of Animal Bioassays Published Through December 1982," *Environmental Health Perspectives* 67 (1986):161–200. L. Gold, T. Slone, G. Backman, R. Magaw, M. Da Costa, and B. Ames, "Second Chronological Supplement to the Carcinogenic Potency Database: Standardized Results of Animal Bioassays Published Through December 1984 and by the National Toxicology Program Through May 1986," *Environmental Health Perspectives* 74 (1987):237–329. L. Gold, T. Slone, G. Backman, S. Eisenberg, M. Da Costa, M. Wong, N. Manley, L. Rohrback, and B. Ames, "Third Chronological Supplement to the Carcinogenic Potency Database: Standardized Results of Animal Bioassays Published Through December 1986 and by the National Toxicology Program through June 1987," *Environmental Health Perspectives* 84 (1990):215–85.

163. Gold, Slone, Backman, Eisenberg, et al., "Third Chronological Supplement."

164. D. Krewski, M. Szyszkowicz, and H. Rosenkranz, "Quantitative Factors in Chemical Carcinogenesis: Variations in Carcinogenic Potency," *Regulatory Toxicology and Pharmacology* 12 (1990):13–29.

165. Cranor, *et al.,* "Improving the Regulation of Carcinogens."

166. Ibid.

167. For 3 of the 74 compounds considered, both ingestion and inhalation potencies are available for comparison.

168. William Pease, Lauren Zeise, and Alex Kelter, "Risk Assessment for Carcinogens under California's Proposition 65," *Risk Analysis* 10(2)(1990):255–71.

169. Cranor, *et al.,* "Improving the Regulation of Carcinogens."

170. Cranor *et al.,* "Improving the Regulatory Control of Carcinogens"; for data selection procedures see the California Environmental Protection Agency, Office of Health Hazard Assessment, Reproductive and Cancer Hazard Assessment Section, "Expected Cancer Potency Values and Proposed Regulatory Levels for Certain Proposition 65 Carcinogens," April 1992.

171. Zeise, "Issues in Risk Assessment," pp. 142–44.

172. The formulas to describe the total social costs of conventional and expedited potency assessments are the following:

Total Social Costs of Conventional Risk Assessments =

$$TC_{CRA} = (N_{MJO} \times C_{MJO}) + (N_{FN} \times C_{FN}),$$

where

N_{MJO} = Number of carcinogens *over*regulated in major way

C_{MJO} = Cost of major overregulation = .9 C_{FP}
C_{FP} = Cost of a regulatory false positive
N_{FN} = Number of carcinogens that are not regulated as such
C_{FN} = Cost of a regulatory false negative.

The first data point for conventional risk assessment at the initial 10:1 ratio is
= (.06 × 78 × .9 × C_{FP}) + (291 × Cost $_{FN}$)
= (.06 × 78 × .9 × $1,000,000) + (291 × 10,000,000)

Total Social Costs of Expedited Risk Assessments =
TC_{ERA} = (N_{MJO} × C_{MJO}) + (N_{mo} × C_{mo}) + (N_{mu} × C_{mu}) + (N_{MJU} × C_{MJU})

where
N_{mu} = Number of carcinogens *under*regulated in minor way
C_{mu} = Cost of minor underregulation = .45 C_{FN}
N_{MJU} = Number of carcinogens *under*regulated in a major way
C_{MJU} = Cost of major underregulation = .9 C_{FN}
N_{mo} = Number of carcinogens *over*regulated in a minor way
C_{mo} – Cost of minor overregulation = .45 C_{FP}
N_{MJO} = Number of carcinogens *over*regulated in major way
C_{MJO} = Cost of major overregulation = .9 C_{FP}

The first data point for the LMS default model scenario at the 10:1 cost ratio on the horizontal axis is = (.06 × 369 × .90 × $1,000,000) + [(.0 × 369 × .9 × $1,000,000) + (.115 × 369 × .45 × $1,000,000) + (.026 × 369 × .90 × $10,000,000) + (.038 × 369 × .45 × $10,000,000)].

The error rate in the first term of the equation is the same as the error rate in the first term of the equation for conventional risk assessments, to reflect that there may be some overestimation of risks on both methods compared to the actual risks in the world. The remainder of the equation describes the accuracy of the LMS potency assessments compared with conventional potency assessments done by the California Department of Health services. The remainder of the data points for conventional and LMS assessments can be generated by appropriate changes in the costs of regulatory false negatives and false positives, and major and minor underregulation and overregulation. Cost figures for the TD_{50} procedure are generated by changing the error rate percentages, plus appropriate modifications in the cost ratios.

173. Lave and Omenn "Cost Effectiveness," and Olson, "The Search for a Safe Environment," Talbot Page, "A Framework for Unreasonable Risk in the Toxic Substances Control Act (TSCA)," in *Annals of the New York Academy of Science: Management of Assessed Risk for Carcinogens,* ed. W. J. Nicholson (New York: New York Academy of Science, 1981), pp. 145–66.

174. Page, in "A Framework for Unreasonable Risk," develops a general framework for minimizing the costs of mistakes and the costs of testing. This point is developed further in Chapter 5.

175. Lave and Omenn, "Cost Effectiveness"; Lave, "A Model for Selecting Short-Term Tests"; Lave and Omenn, "Screening Toxic Chemicals."

176. 5 U.S.C. § 706 (1985);
"The reviewing court shall—
"1) compel agency action unlawfully withheld or unreasonably delayed; and
"2) hold unlawful and set aside agency action, findings, and conclusions found to be—
"A) arbitrary, capricious, an abuse of discretion, or otherwise not in accordance with law;
"B) contrary to constitutional right, power, privilege or immunity;
"C) in excess of statutory jurisdiction, authority, or limitations, or short of statutory right;
"D) without observance of procedure required by law;

"E) unsupported by substantial evidence in a case subject to sections 556 and 557 of this title or otherwise reviewed on the record of an agency hearing provided by statute; or

"F) unwarranted by the facts to the extent that the facts are subject to trial de novo by the reviewing court.

"In making the foregoing determinations, the court shall review the whole record or those parts of it cited by a party, and due account shall be taken of the rule of prejudicial error."

177. Ibid.

178. Zygmunt J. B. Plater and William Lund Norine, "Through the Looking Glass of Eminent Domain: Exploring the 'Arbitrary and Capricious' Test and Substantive Rationality Review of Governmental Decisions," *Boston University Journal of Environmental Affairs* 16 (1989):661, 717.

179. Ibid., p. 717.

180. Ibid.

181. Ibid.

182. Ibid., p. 718 (emphasis added).

183. *Universal Camera Corp. v. National Labor Rel. Bd.*, 340 U.S. 474, 488 (1951) at 487.

184. *Motor Vehicle Mfrs. Assn. v. State Farm Mut.*, 463 U.S. 29, 43 (1982).

185. Ibid.

186. For intrusive and deferential conceptions, respectively, of the "substantial evidence" standard applied to the OSH Act for cases decided shortly after the *Benzene* case, compare *Texas Independent Ginners Ass'n. v. Marshall*, 630 F.2d 398, 404 (5th Cir. 1980) with *United Steelworkers v. Marshall*, 647 F.2d 1189, 1207 (D.C. Cir. 1980). Part of this difference may be accounted for by the fact that *Texas Independent Ginners* was decided in the Court of Appeals for the 5th Circuit, the same court that decided the *Gulf South* case.

187. Whether court deference will result on better protection for human health and the environment depends upon what agencies do with their discretion. If they are better "captured" by the industries they seek to regulate, this is not a desirable outcome. If they use their discretion to design health-protective expedited procedures, we will be better protected. I recommend agencies develop such procedures.

188. Richard J. Pierce, Jr., "Two Problems in Administrative Law: Political Polarity on the District of Columbia Circuit and Judicial Deterrence of Agency Rulemaking," *Duke Law Journal*, 1988, p. 300.

189. Ibid., p. 302.

190. Ibid., p. 314, citing J. Mashaw, "Prodelegation: Why Administrators Should Make Political Decisions," *Journal of Law Economics and Organizations.* 1 (1985):82; T. McGarity, "Regulatory Analysis and Regulatory Reform," *Texas Law Review* 65 (1987):1243; P. L. Strauss, "One Hundred Fifty Cases per Year: Some Implications of the Supreme Court's Limited Resources for Judicial Review of Agency Action," *Columbia Law Review* 87 (1987):1093.

191. Pierce, "Two Problems," p. 316.

192. Alfred C. Aman, Jr., "Administrative Law in a Global Era: Progress, Deregulation, Change, and the Rise of the Administrative Presidency," *Cornell Law Review* 73 (1988):1101, esp. pp. 1101–1242.

193. Whether or not judicial deference for agency rulemaking is desirable or not depends upon what the agencies do and one's view of correct environmental health policy. Regulations that gutted health protections would offend environmentalists but might please the regulated industries. Thus, judicial deference to such actions cannot be *fully* evaluated apart from the policy outcomes. However, for purposes of permitting agencies to experiment with expedited and novel risk assessments, judicial deference appears to be a good thing.

194. Dwyer, "Overregulation," pp. 719, 731; Paolo F. Ricci, Louis A. Cox, and John P. Dwyer, "Acceptable Cancer Risks: Probabilities and Beyond," *Journal of the Air Pollution Control Association,* 39 (1989):1046, 1051; "Begging to Defer: OSHA and the Problem of Interpretive Authority," *Minnesota Law Review 73* (1989):1336, 1346, 1356 (note).

195. Dwyer, "Overregulation," p. 731.

196. 462 U.S. 87.

197. Ibid., at 103, referencing both the plurality and dissenting opinions in *Industrial Union Dept., AFL-CIO v. American Petroleum Institute,* at 607, 656 (plurality opinion), at 705–6 (Justice Marshall, dissenting).

198. *Chevron U.S.A. Inc. v. Natural Resources Defense Council, Inc.,* 467 U.S. 837 (1983).

199. Ibid., at 865.

200. John P. Dwyer, University of California, Berkeley, Boalt Law School, personal communication, June 16, 1990.

201. *ASARCO, Inc. v. OSHA,* 746 F.2d 483, 490–95 (9th Cir. 1984).

202. Ibid., at 492–95.

203. *Public Citizen Health Research Group v. Tyson,* 796 F.2d 1479 (D.C. Cir. 1986).

204. Ibid., at 1495, 1496–98, 1498.

205. Ibid., at 1499 (emphasis added).

206. *Building and Constr. Trades Dept., AFL-CIO v. Brock,* 838 F.2d 1258 (D.C. Cir. 1988).

207. Ibid., at 1266.

208. Ibid., at 1266–67, quoting *Industrial Union Department, AFL-CIO v. American Petroleum Institute,* 448 U.S. 607 (1980) at 656.

209. *AFL-CIO v. Brock* at 1267 (emphasis in original).

CHAPTER 5

1. A. E. Ades, "Evaluating Screening Tests and Screening Programs," *Archives of Disease in Childhood* 65 (1990):793.

2. Ibid.

3. Talbot Page, "A Framework for Unreasonable Risk in the Toxic Substances Control Act (TSCA)," in *Annals of the New York Academy of Science: Management of Assessed Risk for Carcinogens,* ed. W. J. Nicholson (New York: New York Academy of Sciences, 1981), pp. 145–66, discusses and develops such a cost-minimizing strategy.

4. Lester B. Lave, Fanny K. Ennever, Herbert S. Rosenkranz, and Gilbert S. Omenn, "Information Value of the Animal Bioassay," *Nature* 336 (1988):631–33. Lave also bases his argument on a cost minimization strategy.

5. A common tort law principle endorsed by the courts suggesting this point is that if the courts have to choose between failing to compensate an injured, innocent plaintiff (a false negative) and wrongly imposing compensation costs on a faultily acting agent whose actions may be causally "innocent" (for they did not cause plaintiff's harm) (a false positive), the courts prefer to make the second mistake rather than the first. (*Sindell v. Abbott Laboratories,* 76 Cal. 3d 588 (1980): *Summers v. Tice,* 33 Cal. 2d 80. 199 (1948), p. 201.)

Of course, this principle does not settle the issue, but does suggest that courts, *in this class of cases,* prefer more strongly to avoid false negatives than to avoid false positives. The burdens of proof ordinarily placed on plaintiffs argue in the other direction.

6. See Mark Sagoff, *The Economy of the Earth* (New York: Cambridge University Press, 1988), and Douglas MacLean, "Social Values and the Distribution of Risk," in *Values*

at Risk, ed. D. MacLean (Totowa, NJ: Rowman and Allenheld, 1986), pp. 75–93, who express such views.

7. An extreme version of this problem is posed by Pascal's Wager. Suppose one has no evidence about the existence of God but must decide whether to believe that God exists. Either God exists or not and either one believes God exists or not; there are four possibilities. If one believes God exists and God does not, then one will have a somewhat less interesting and exciting life than one might have had. But if God exists and one believes that God exists (and acts accordingly), one will be rewarded eternally. However, if one believes that God does not exist but God exists, one will suffer eternally. And if one believes God does not exist and God does not, then one will have the rewards of a somewhat more interesting and exciting life than otherwise. Pascal argues that one choice dominates; one should believe God exists.

This argument is not sound, but it illustrates decision making under almost complete uncertainty. In this circumstance, where the facts are uncertain, the choice is guided by what is at stake, by the values secured as a result of each of the choices.

For a decision in administrative law in which a court notes the importance of values (policy) guiding choices when considerable factual uncertainty obtains, see *Industrial Union Department, AFL-CIO v. Hudson* (499 F.2d 467 (1974) at 474.

> Decision making must in that circumstance [when insufficient data is present to make a fully informed factual decision] depend to a greater extent upon policy judgments and less upon purely factual analysis.

8. I borrow this term from T. M. Scanlon, "Preference and Urgency," *Journal Philosophy* 77 (1975):655–59.

9. Annette Baier, "Poisoning the Wells," in *Values at Risk,* ed. D. MacLean (Totowa, N.J. Rowman & Allenheld, 1986), pp. 49–75.

10. A clear understanding of the implications of principles is not the only consideration bearing on a choice between them, of course, but it is an important first step.

11. John Rawls, *A Theory of Justice* (Cambridge, Mass: Harvard University Press, 1971), p. 3.

12. Anecdotal evidence indicates that in many or even most cases persons' views on risk assessment models are dictated by the normative outcome they think desirable, not vice versa.

13. This is explained in the Introduction.

14. 29 U.S.C., § 655 (1976).

15. *American Iron & Steel, Inst. v. OSHA,* 577 F.2 825, 831 (1978).

16. *Society of Plastics Industry, Inc. v. OSHA,* 509 F.2d 1301, 1306 (1975).

17. *Industrial Union Dept., AFL-CIO v. Hodgson,* 499 F.2d 467, 471 (1974).

18. "Occupational Exposure to Asbestos," *NIOSH Criteria Document* (1972), at Ch. III, p. 7.

19. *American Textile Mfrs. Inst. v Donovan,* 452 U.S. 490, 498 (1980).

20. ASARCO, Inc. v. OSHA, 746 F.2d 483 (9th Cir., 1984).

21. *Public Citizen Health Research Group v. Tyson,* 796 F.2d 1479, 1493 (D.C. Cir. 1986).

22. This is the principle from the law, "volenti non fit inuria." *Black's Law Dictionary,* Rev. 4th Ed. (St. Paul, Minn: West Publishing Company, 1968), p. 1747.

23. Thomas Hobson (circa 1631), an English liveryman, required every customer to take the horse that stood nearest the door, thus providing an apparent freedom of choice where there was no real alternative.

24. Norman Daniels, "Does OSHA Regulate Too Much?" in *Just Health Care* (New York: Cambridge University Press, 1985), discusses this issue.

25. 29 U.S.C., § 651–678 (1976).

26. 29 U.S.C., § 655 (b)(5)(1976) (emphasis added).

27. Technological feasibility is also important, even though I do not discuss it. Under the OSH Act OSHA is not obliged to achieve the no-material-impairment-of-health goal if it is not technologically feasible. However, whether such goals are technologically possible depends not only on what devices or strategies are currently available, but also upon what companies have been urged or given incentives to develop. Thus, governments and societies as a whole can also be judged on the extent to which they have encouraged such technologies. If there are no devices available to protect against toxic substances, a society may have been remiss and at fault for not placing a higher priority on them.

28. In "Does OSHA Regulate Too Much?" Norman Daniels addresses a libertarian objection to OSHA's regulations: Why reduce toxic exposures to the lowest level *technologically feasible* when employees might voluntarily subject themselves to higher risks for hazard pay?

29. *Industrial Union Dept., AFL-CIO v. Hodgson*, 162 U.S. App. D.C. at 342, 499 F.2d at 467, 477–78 (D.C. Cir. 1974).

30. *Industrial Union Dept., AFL-CIO v. Marshall*, 199 U.S. App. D.C. 54, 617 F.2d 636 (1979) at 661. All references to 617 F.2d 636 (1979).

31. Ibid., at 661–62.

32. Ibid., at 661.

33. It is also likely that more stringent regulations would disproportionately affect smaller companies, perhaps forcing some to close.

34. *Industrial Union Dept., AFL-CIO v. Marshall*, at 665.

35. American Textile Mfgs. Inst. v. Donovan, 452 U.S. 490 at 509, note 29 (1981).

36. *American Textile Mfrs, Inst. v. Donovan*, at 506–22 (1981).

37. See, e.g., *United Steelworkers, AFL-CIO v. Brennan*, 530 F.2d 109, 122 (3d, Cir. 1975); *Industrial Union Dept., AFL-CIO v. Hodgson*, 499 F.2d 467, 478 (D.C. Cir. 1974), as well as H. Latin, "The Feasibility of Occupational Health Standards: An Essay on Legal Decision-making Under Uncertainty," 78 *Northwestern University Law Review* 78(1983) 583, 588 who makes a similar point.

38. *Universal Camera Corp. v. National Labor Relations Bd.*, 340 U.S. 474, 488 (1951).

39. See Richard Brandt, *A Theory of the Good and the Right* (Oxford: Clarendon Press, 1979), pp. 246–65, for some discussion of these issues.

40. We should note that good health is not a universal means, for there are a number of ends that can be pursued even though one's health is not good or excellent.

41. Although this is a difficulty for utilitarianism, some theorists explicitly argue for *rights* as part of utilitarianism and almost all try to give some argument for distributive considerations. David Lyons, "Human Rights and General Welfare," *Philosophy and Public Affairs* (6)(2):113–29 (1977).

42. The "unreasonable risk" criterion of several environmental health statutes (e.g., provisions of the Federal Insecticide, Fungicide and Rodenticide Act and of the Toxic Substances Control Act) resembles utilitarian principles in that they permit economic and other benefits to be weighed against any health harms. These are contrasted with statutes that protect health "with an ample margin of safety" or "no-risk" statutes such as the FDCA's Delaney Clause, discussed in Chapter 4.

43. Utilitarians would respond that the law of diminishing marginal utilities provides some additional distributions to those who are worst off, since they likely can obtain greater utility from a particular increment of resources than can someone who is much better off.

44. L. Tomatis, "Environmental Cancer Risk Factors: A Review," *Acta Oncologica* 27(1988):166, and L. Tomatis, A. Aito, N. E. Day, E. Heseltine, J. Kaldor, A. B. Miller, D. M. Paskin, & E. Biboli, *Cancer: Causes, Occurrence and Control* (Lyon, France: International Agency for Research on Cancer, 1990) p. 4.

45. J. J. C. Smart and Bernard Williams, *Utilitarianism: For and Against* (Cambridge: Cambridge University Press, 1973), p. 37.

46. Rawls, *A Theory of Justice*, p. 27.

47. N. Rescher, *Distributive Justice* (New York: Bobbs-Merrill, 1966), Introduction, shows the need for welfare floors.

48. See S. Scheffler, *The Rejection of Consequentialism: A Philosophical Investigation of the Considerations Underlying Rural Moral Conceptions* (New York: Oxford University Press, 1982) and Gregory Kavka "The Numbers Should Count," *Philosophical Studies,* Vol. 36, pp. 285–94 (1979) for some discussion of these issues.

49. Scheffler, *The Rejection of Consequentialism,* p. 29.

50. Rawls, *A Theory of Justice,* p. 554.

51. Scheffler, *The Rejection of Consequentialism,* p. 28.

52. Norman Daniels, "Health Care Needs and Distributive Justice," *Philosophy and Public Affairs* 10 (1981):146–79.

53. Utilitarians and their economic offspring, economists, would argue that the law of diminishing marginal utility would go a long way toward securing such *distributive* results. While there is some truth in this, it does not guarantee such results and is ultimately an unsatisfactory answer.

54. Daniels, "Health Care Needs," pp. 146–79.

55. Rawls, *A Theory of Justice,* p. 3.

56. Daniels, "Health Care Needs," p. 166.

57. Norman Daniels, *Just Health Care* (New York: Cambridge University Press, 1985), p. 43.

58. Ibid., p. 46.

59. Ibid., p. 47 (emphasis added).

60. Ibid., p. 33.

61. Ibid.

62. Daniels, *Just Health Care,* p. 33, and Rawls, *A Theory of Justice,* pp. 83–89.

63. Daniels *Just Health Care,* p. 35.

64. Ibid., p. 33.

65. Daniels, *Just Health Care,* p. 35; Daniels, *Health Care Needs,* p. 166.

66. Ibid., pp. 167–68. A second layer of institutions corrects "departures from the idealization" by attempting to restore normal functioning, and a third layer "attempts, where feasible, to maintain persons in a way that is close as possible to the idealization." A fourth layer "involves health care and related social services for those who can in no way be brought closer to the idealization."

67. Ibid., p. 160.

68. Ibid., p. 166.

69. Daniels, *Just Health Care,* p. 104.

70. Buchanan, "The Right to a Decent Minimum of Health Care," *Philosophy and Public Affairs,* p. 63. In *Just Health Care* Daniels acknowledges these other points but thinks that the concern about disease undermining one's normal opportunity range is the more fundamental one (pp. 36–56).

71. This egalitarian aim cannot always be ensured for health protections, for what is to be distributed *equally* on the Daniels-Rawls' view are opportunities to pursue one's life plan, and the overall equal distribution of opportunities may in particular cases necessitate the unequal

distribution of health care. However, the justice principle permits fewer reasons to depart from the ideal of equality than do typical utilitarian theories.

72. In short, working at jobs where consciously taking risks was part of what made one's life worth living. The risks from chemical carcinogens do not seem to be included in this class of jobs, since they are difficult or impossible to detect, exposure to them has little of the romance that being a stunt double or high-rise construction worker might be thought to have.

73. Rawls, *A Theory of Justice*, 302–3. According to Rawls's special conception of justice, the fair equality of opportunity part of the second principle of justice must be fully satisfied before one considers fulfilling the difference principle, which is concerned with the distribution of wealth and power.

74. See Norman Daniels, *Reading Rawls* (New York: Basic Books, 1976) for discussions of his theory.

75. The extent of health protections Daniels indicates is given by the extent to which diseases caused by the absence of such protections would curtail "an individual's share of the normal opportunity range." *Just Health Care*, p. 35.

76. Rawls, *A Theory of Justice*, p. 301.

77. Ibid.

78. T. M. Scanlon, "Rawls' Theory of Justice," *Pennsylvania Law Review* 121 (1973):1020–69, esp. pp. 1028–29, 1051–53; Scheffler, *The Rejection of Consequentialism*.

79. I have not documented this view, although I think it quite easy to do. Many hours of observation at congressional hearings, interviews with administrative agency staff during 1985–86 when I was a congressional fellow, and numerous conferences attended by industry, governmental, and environmental representatives as well as by academics from other fields indicate to me that a large group of people assesses social and governmental decisions from a normative perspective much like utilitarianism.

80. Talbot Page, "A Generic View of Toxic Chemical and Similar Risks," *Ecology Law Quarterly*, 7 (1978):207,109.

81. In arguing that particular legal institutions should give greater weight to avoiding false negatives than to avoiding false positives, I relied intuitively on a somewhat vague distributive view. There is, however, one way in which I also conceded something to utilitarianism—I assumed that the social costs of false negatives and false positives are comparable, that they can be compared on the same scale of value. Thus, for example, one argument from Chapter 4 assumed both FNs and FPs could be assigned dollar values and that the ratio of these costs could vary from 10:1 down to 1:10. This may concede too much in the direction of utilitarian kinds of tradeoffs, for it assumes comparability, as well as the summing of minor benefits that could eventually outweigh the costs of a FN. (Nonetheless, even that evaluation of risk assessment argues for reform!)

Firm reliance on a distributively sensitive theory might well limit such tradeoffs. Even if a distributively sensitive theory permitted some comparability, and at a certain point if a large enough number of minor benefits could outweigh severe costs to a few, the cost of FNs would be much greater than the costs of FPs—to oversimplify, a ratio 10:1 or even greater. Thus, even though many of the arguments offered throughout are consistent with the less defensible utilitarian view, they are, I believe, persuasive, and they are even stronger if a more distributively sensitive theory were used to assign these ratios.

82. Another possible view that might satisfy the requirements of a distributively sensitive consequentialism might be a modified rule utilitarian theory similar to one proposed by Richard Brandt in *A Theory of the Good and the Right*, provided that concerns about a better conception of the human good could also be met. The idea would be to choose rules for governing workplace and environmental health protections such that the rules as a group would be justified by utilitarian considerations. One would choose the set of compatible rules

that would likely produce as much or more human good as would adopting any alternative set of rules for providing environmental health protections. In such choice conditions it is unlikely that people would subscribe to rules which permitted severe costs to some to be outweighed by minor benefits to a large number of people. The specter of such rules would make everyone feel sufficiently insecure that they would not be willing to live in accordance with them. Moreover, such rules might substantially interfere with one's opportunities. (Whether such a highly modified utilitarian principle would be counted any longer as utilitarian would be open to some debate.) (Craig Ihara suggested this point.)

A decision procedure that might yield distributively sensitive principles is a contractarian form of argument with decision makers in the choice position taking into account not only welfare effects but also opportunities similar to those Daniels suggests. An important feature of contracarian theories is how the principles considered for adoption in the choice position would affect *individual* persons. This is contrasted with many versions of utilitarianism in which the concern is with changes in aggregate well-being. Contractarian views also tend to be motivated by a desire to "justify one's actions to others on grounds they could not reasonably reject," as Scanlon indicates ("Contractualism and Utilitarianism," p. 116).

APPENDIX A

Uncertainties in Carcinogen Risk Assessments

1. Hazard Identification—Classifying which animal carcinogens are human carcinogens.
2. Use of most sensitive sex/strain/tumor site in typical rodent studies vs. average sensitivity in a species. Magnitude of uncertainty is a factor of 1–100.
3. Use of human data vs. most sensitive animal sex/species/strain/tumor site. Magnitude of uncertainty is a factor of 1–1000.
4. Count benign tumors or not.
5. Use of linear vs. nonlinear extrapolation models. Magnitude of uncertainty is a factor of 1–1,000,000.
6. Use of upper 95% confidence limits vs. plausible upper bounds vs. maximum likelihood estimates. Magnitude of uncertainty is a factor of 1–10.
7. Use of different interspecies scaling factors: surface area (body weight$^{2/3}$) vs. body weight$^{3/4}$ vs. body weight. Magnitude of uncertainty is a factor of 1–13.
8. Use of genotoxic vs. epigenetic distinction for regulatory purposes. Magnitude of uncertainty is a factor of 1–10.
9. Use of pharmacokinetic information in risk assessment: dose of substance at exchange boundary vs. dose at target organ. Magnitude of uncertainty is a factor of 1–10.
10. Risks estimated for total population vs. target population. Magnitude of uncertainty is a factor of 1–10,000.
11. Use of different *de minimus* risk thresholds to trigger regulatory action. The threshold may be 10^{-4}, 10^{-5}, or 10^{-6}.
12. Adding vs. not adding theoretical risks for different substances. Magnitude of uncertainty is a factor of 1–100.
13. Consideration of potential synergisms or antagonisms with other carcinogens or promoters. Magnitude of uncertainty is unknown but may be a factor of 1–1000.

Information is based on data from Robert Brown of the FDA, and other sources. Compiled by Kenneth Dickey, University of California, Riverside.

From Carl Cranor, "An Overview of Risk Assessment," in *Proceedings: Pesticides and Other Toxics: Assessing Their Risks,* ed. Janet White (Riverside: University of California, College of Natural and Agricultural Sciences, 1990), p. 83.

APPENDIX B

Cancer Potency Estimates of CDHS and EPA [(mg/kg-day)$^{-1}$]

Substance	CDHS Range of Potency Values	CDHS Recommended Potency	EPA Potency	CDHS/EPA Potency Ratio
Acetaldehyde	In process	In process	0.0077	—
Acrylonitrile				
Ingestion	—	1.0	0.54	2[*,b]
Inhalation	—	1.0	0.24	4[a]
Aflatoxin	30–26,326	98	Not available	
Aldrin	—	17	16	1
Asbestos				
Ingestion	Postponed		7×10^{-4}/fiber[†]	
Inhalation	10^{-8}–10^{-7}	10^{-7}/PCM fiber	2×10^{-8}/fiber	5[a]
Benzene	0.04–0.26	0.1	0.029	3.4[a,b]
Benzidine	7–27,000	500	234	1.5[a]
Benzo(a)pyrene	In process	In process	[11.5][‡]	
Beryllium oxide				
Ingestion	Postponed	Postponed	4.3	—
Inhalation	In process	In process	7	—
Beryllium sulfate				
Ingestion	Postponed	Postponed	4.3	—
Inhalation	In process	In process	300	—
Bis(chloroethyl)ether	0.5–2.5	2.5	1.14	2.2
Bis(chloromethyl)ether	26–240	46	220[9,300][‡]	0.2
1,3-Butadiene	In process	In process	1.8	—
Cadmium				
Ingestion	Postponed	Postponed	—	—
Inhalation		14.6	6.1	2.4[a]
Carbon tetrachloride				
Ingestion	0.01–17	0.18	0.13	1.4[d]
Inhalation	0.01–17	0.15	0.13	1.2
Chlordane	—	—	1.3	—
Chloroform				
Ingestion	In process	In process	0.0061	—
Inhalation	In process	In process	0.0081	—
Chromium VI				
Ingestion	Postponed	Postponed	—	—
Inhalation	60–3,200	500	41	12[a]
Coke oven emissions	—	—	2.16	—

From Lauren Zeise, "Issues in Risk Assessment: California Department of Health Services," in *Proceedings: Pesticides and Other Toxics: Assessing Their Risks,* ed. Janet White (Riverside: University of California, College of Natural and Agricultural Sciences, 1990), pp. 139–41.

Substance	CDHS Range of Potency Values	CDHS Recommended Potency	EPA Potency	CDHS/EPA Potency Ratio
DBCP	0.2–100	7	Not available	—
DDT	—	0.34	0.34	1
1,4-Dichlorobenzene	—	7	Not available	—
3–3-Dichlorobenzene	0.008–1.2	1.2	1.69	0.7
Dichloromethane				
Ingestion	In process	In process	0.0075	—
Inhalation	In process	In process	0.014	—
1,3-Dichloropropene	—	0.18	0.18	1
Dieldrin	—	16	20	0.8
2,4-Dinitrotoluene	—	—	0.31	—
1,4-Dioxane	0.001–0.027	0.027	0.011	2.4
Epichlorohydrin				
Ingestion	0.03–2.1	0.08	0.0099	8[c]
Inhalation	0.03–2.1	0.08	0.0042	19[c,d]
Ethylene dibromide				
Ingestion	0.7–100	3.6	41	0.09[c]
Inhalation	0.2–4	0.25	41	0.006[c]
Ethylene dichloride	0.01–1.9	0.07	0.091	0.77
Ethylene oxide	0.01–0.35	0.35	0.35	1.0
Heptachlor	—	—	4.5	—
Heptachlor epoxide	—	—	9.1	—
Hexachlorobenzene	0.08–1.8	1.8	1.67	1.1
Hexachlorocyclohexane (technical grade)	0.8–6.7	4	1.8	2.2
Nickel refinery dust	In process	In process	0.84	—
Nickel subsulfide	In process	In process	1.7	—
N-nitrosodibutylamine	4–40	10.8	5.43	—
N-nitrosodiethylamine	12–1,100	36	150	0.2[c]
N-nitrosodimethyl-amine	12–100	16	51	0.3[c]
N-nitrosodiphenylamine	0.003–0.06	0.009	0.00492	1.8
N-nitrosopyrrolidine	—	—	2.13	—
N-nitroso-N-ethylurea	29–150	27	32.9	0.8[c]
N-nitroso-N-methylurea	106–1,800	124	302.6	0.4
N-nitrosodipropylamine	—	7	7	1
PBBs	—	30	Not available	—
PCBs(>60%)	4.2–5.0	5	7.7	0.6
TCDD	In process	In process	156,000	—
Tetrachloroethylene	In process	In process	0.051	—
Toxaphene	0.08–3	1.2	1.13	1.1
Trichlorethylene				
Ingestion	In process	In process	0.011	—
Inhalation	In process	In process	0.013	—
2,4,6-Trichlorophenol	0.008–0.47	0.07	[0.0199]‡	3.5

(*Continued*)

Substance	CDHS Range of Potency Values	CDHS Recommended Potency	EPA Potency	CDHS/EPA Potency Ratio
Unleaded gasoline	—	0.0035	0.0035	1
Urethane	0.01–12	1	Not available	—
Vinyl chloride	In process	In process	2.3	—

*Superscripts a–e refer to differences between EPA and CDHS numbers primarily due to (a) CDHS use of upper 95% confidence bound estimate, EPA's use of maximum likelihood estimate; (b) CDHS used upper bound estimate from human study; EPA, the geometric mean of upper bound estimates from four animal experiments; (c) survival corrections; (d) use of different studies; and (e) EPA documentation not available.

†Environmental Protection Agency (1985). *Federal Register* 50(219):45,929–46,963.

‡Potency estimate retracted by EPA.

Data also appeared in W. S. Pease, L. Zeise, A. Kelter (1990). Risk assessment for carcinogens under California's Proposition 65. *Risk Analysis,* Vol. 10, No. 2.

APPENDIX C

Relative Risk as a Function of Alpha and Beta Values

A. Relative Risk When Background Disease Rate is 8/10,000

	Alpha	Beta	Bkgd.Dis.Rate	Rel.Risk	Sample
1	0.05	0.05	8.0E-04	8.9	2150
2		0.10	8.0E-04	7.5	2150
3		0.15	8.0E-04	6.7	2150
4		0.20	8.0E-04	6.0	2150
5		0.25	8.0E-04	5.5	2150
6		0.30	8.0E-04	5.1	2150
7		0.35	8.0E-04	4.7	2150
8		0.40	8.0E-04	4.3	2150
9		0.45	8.0E-04	4.0	2150
10		0.49	8.0E-04	3.8	2150
11	0.10	0.05	8.0E-04	7.5	2150
12		0.10	8.0E-04	6.2	2150
13		0.15	8.0E-04	5.5	2150
14		0.20	8.0E-04	4.9	2150
15		0.25	8.0E-04	4.5	2150
16		0.30	8.0E-04	4.1	2150
17		0.35	8.0E-04	3.8	2150
18		0.40	8.0E-04	3.5	2150
19		0.45	8.0E-04	3.2	2150
20		0.49	8.0E-04	3.0	2150
21	0.15	0.05	8.0E-04	6.7	2150
22		0.10	8.0E-04	5.5	2150
23		0.15	8.0E-04	4.8	2150
24		0.20	8.0E-04	4.3	2150
25		0.25	8.0E-04	3.9	2150
26		0.30	8.0E-04	3.6	2150
27		0.35	8.0E-04	3.2	2150
28		0.40	8.0E-04	3.0	2150
29		0.45	8.0E-04	2.7	2150
30	0.20	0.05	8.0E-04	6.0	2150
31		0.10	8.0E-04	4.9	2150
32		0.15	8.0E-04	4.3	2150
33		0.20	8.0E-04	3.8	2150
34		0.25	8.0E-04	3.4	2150
35		0.30	8.0E-04	3.1	2150
36		0.35	8.0E-04	2.8	2150
37		0.40	8.0E-04	2.6	2150
38		0.45	8.0E-04	2.4	2150

(*Continued*)

	Alpha	Beta	Bkgd.Dis.Rate	Rel.Risk	Sample
39	0.25	0.05	8.0E-04	5.5	2150
40		0.10	8.0E-04	4.5	2150
41		0.15	8.0E-04	3.9	2150
42		0.20	8.0E-04	3.4	2150
43		0.25	8.0E-04	3.1	2150
44		0.30	8.0E-04	2.8	2150
45		0.35	8.0E-04	2.5	2150
46		0.40	8.0E-04	2.3	2150
47		0.45	8.0E-04	2.1	2150
48	0.30	0.05	8.0E-04	5.1	2150
49		0.10	8.0E-04	4.1	2150
50		0.15	8.0E-04	3.5	2150
51		0.20	8.0E-04	3.1	2150
52		0.25	8.0E-04	2.8	2150
53		0.30	8.0E-04	2.5	2150
54		0.35	8.0E-04	2.2	2150
55		0.40	8.0E-04	2.0	2150
56		0.45	8.0E-04	1.8	2150
57	0.33	0.05	8.0E-04	4.9	2150
58		0.10	8.0E-04	3.9	2150
59		0.15	8.0E-04	3.4	2150
60		0.20	8.0E-04	2.9	2150
61		0.25	8.0E-04	2.6	2150
62		0.30	8.0E-04	2.4	2150
63		0.35	8.0E-04	2.1	2150
64		0.40	8.0E-04	1.9	2150
65		0.45	8.0E-04	1.7	2150
66	0.40	0.05	8.0E-04	4.3	2150
67		0.10	8.0E-04	3.5	2150
68		0.15	8.0E-04	3.0	2150
69		0.20	8.0E-04	2.6	2150
70		0.25	8.0E-04	2.3	2150
71		0.30	8.0E-04	2.0	2150
72		0.35	8.0E-04	1.8	2150
73		0.40	8.0E-04	1.6	2150
74		0.45	8.0E-04	1.5	2150
75	0.45	0.05	8.0E-04	4.0	2150
76		0.10	8.0E-04	3.2	2150
77		0.15	8.0E-04	2.7	2150
78		0.20	8.0E-04	2.4	2150
79		0.25	8.0E-04	2.1	2150
80		0.30	8.0E-04	1.8	2150
81		0.35	8.0E-04	1.6	2150
82		0.40	8.0E-04	1.5	2150
83		0.45	8.0E-04	1.3	2150

B. Relative Risk When Background Disease Rate is 8/100,000

	Alpha	Beta	Bkgd.Dis.Rate	Rel.Risk	Sample
1	0.05	0.05	8.0E-05	65.9	2150
2		0.10	8.0E-05	52.6	2150
3		0.15	8.0E-05	44.8	2150
4		0.20	8.0E-05	38.8	2150
5		0.25	8.0E-05	34.1	2150
6		0.30	8.0E-05	30.4	2150
7		0.35	8.0E-05	26.9	2150
8		0.40	8.0E-05	23.8	2150
9		0.45	8.0E-05	21.2	2150
10	0.10	0.05	8.0E-05	52.6	2150
11		0.10	8.0E-05	40.9	2150
12		0.15	8.0E-05	34.1	2150
13		0.20	8.0E-05	28.9	2150
14		0.25	8.0E-05	24.9	2150
15		0.30	8.0E-05	21.8	2150
16		0.35	8.0E-05	18.9	2150
17		0.40	8.0E-05	16.4	2150
18		0.45	8.0E-05	14.3	2150
19	0.15	0.05	8.0E-05	44.8	2150
20		0.10	8.0E-05	34.1	2150
21		0.15	8.0E-05	28.0	2150
22		0.20	8.0E-05	23.4	2150
23		0.25	8.0E-05	19.8	2150
24		0.30	8.0E-05	17.1	2150
25		0.35	8.0E-05	14.5	2150
26		0.40	8.0E-05	12.4	2150
27		0.45	8.0E-05	10.6	2150
28	0.20	0.05	8.0E-05	38.8	2150
29		0.10	8.0E-05	28.9	2150
30		0.15	8.0E-05	23.4	2150
31		0.20	8.0E-05	19.2	2150
32		0.25	8.0E-05	16.0	2150
33		0.30	8.0E-05	13.6	2150
34		0.35	8.0E-05	11.4	2150
35		0.40	8.0E-05	9.5	2150
36	0.33	0.05	8.0E-05	28.2	2150
37		0.10	8.0E-05	20.0	2150
38		0.15	8.0E-05	15.5	2150
39		0.20	8.0E-05	12.2	2150
40		0.25	8.0E-05	9.8	2150
41		0.30	8.0E-05	8.0	2150
42		0.35	8.0E-05	6.4	2150
43		0.40	8.0E-05	5.1	2150
44		0.45	8.0E-05	4.1	2150
45	0.40	0.05	8.0E-05	23.8	2150
46		0.10	8.0E-05	16.4	2150
47		0.15	8.0E-05	12.4	2150
48		0.20	8.0E-05	9.5	2150

(*Continued*)

	Alpha	Beta	Bkgd.Dis.Rate	Rel.Risk	Sample
49		0.25	8.0E-05	7.4	2150
50		0.30	8.0E-05	6.0	2150
51		0.35	8.0E-05	4.6	2150
52		0.40	8.0E-05	3.6	2150
53		0.45	8.0E-05	2.8	2150
54	0.45	0.05	8.0E-05	21.2	2150
55		0.10	8.0E-05	14.3	2150
56		0.15	8.0E-05	10.6	2150
57		0.20	8.0E-05	8.0	2150
58		0.25	8.0E-05	6.2	2150
59		0.30	8.0E-05	4.8	2150
60		0.35	8.0E-05	3.7	2150
61		0.40	8.0E-05	2.8	2150
62		0.45	8.0E-05	2.1	2150

APPENDIX D

Statutes Authorizing Regulation of Carcinogens

Legislation	Agency	Area of concern
Food, Drug, and Cosmetic Act (1906, 1938, amended 1958, 1960, 1962, 1968)	FDA	Foods, drugs, cosmetics, and medical devices
Federal Insecticide, Fungicide and Rodenticide Act (1948, amended 1972, 1975, 1978)	EPA	Pesticides
Dangerous Cargo Act (1952)	DOT, USCG	Water shipment of toxic materials
Atomic Energy Act (1952)	NRC	Radioactive substances
Federal Hazardous Substances Act (1960, amended 1961)	CPSC	Toxic household products
Federal Meat Inspection Act (1967)	USDA	Food, feed, color additives, pesticide residues
Poultry Products Inspection Act (1970)		
Egg Products Inspection Act		
Occupational Safety and Health Act (1970)	OSHA	Workplace toxic chemicals
Poison Prevention Packaging Act (1970, amended 1977)	CPSC	Packaging of hazardous household products
Clean Air Act (1970, amended 1974, 1977)	EPA	Air pollutants
Hazardous Materials, Transportation Act (1972)	DOT	Transport of hazardous materials
Clean Water Act (formerly Federal Water Control Act) (1972, amended 1977, 1978)	EPA	Water pollutants
Marine Protection, Research and Sancturies Act (1972)	EPA	Ocean dumping
Consumer Product Safety Act (1972, amended 1981)	CPSC	Hazardous consumer products
Lead-based Paint Poison Prevention Act (1973, amended 1976)	CPSC, HHS, HUD	Use of lead paint in federally assisted housing
Safe Drinking Water Act (1976)	EPA	Drinking water contaminants
Resource Conservation and Recovery Act (1976)	EPA	Solid waste
Toxic Substances Control Act (1976)	EPA	Hazardous chemicals not covered by other acts
Federal Mine Safety and Health Act (1977)	DOL, NIOSH	Toxic substances in coal and other mines
Comprehensive Environmental Response, Compensation, and Liability Act (1981)	EPA	Hazardous waste cleanup

From U.S. Congress, Office of Technology Assessment, *Identifying and Regulating Carcinogens* (Washington, D.C.: GPO, 1987), p. 199.

APPENDIX E

Derivation of TD$_{50}$ Potency Values

$$P_c = 1 - e - (q_0 + q_1 \times d + q_2 \times d^2 + \cdots) \tag{1}$$

where P_c = probability of contracting cancer
  q_0 = background cancer rate,
  q_1 = contribution to cancer rate from first stage event, q_2 from second stage event, etc.

$$P_c = 1 - e^{-q_1 \times d} \tag{2}$$

from (1) as a result of simplifying by ignoring background and subsequent stages' contributions to cancer

$$0.50 = 1 - e^{-q_1 \times d} \tag{3}$$

50% cancer rate from TD$_{50}$ experiment

$$e^{-q_1 \times d} = \frac{1}{2} \tag{4}$$

rearrangement of (3)

$$e^{-q_1 \times \text{TD}_{50}} = \frac{1}{2} \tag{5}$$

substitution of TD$_{50}$ dose for d

$$q_1 \times \text{TD}_{50} = \ln 2 \tag{6}$$

natural log of both sides of equation

$$q_1 = \ln 2 / \text{TD}_{50} \tag{7}$$

rearrangement of (6)

Bibliography

Abel, R. (1990). A critique of torts. *UCLA Law Review* 37:785–831.

Ackerman, B. (1980). Lecture notes from environmental law course, Yale Law School.

Ades, A. E. (1990). Evaluating screening tests and screening programs. *Archives of Disease in Childhood* 65:792–95.

Allen, B. C., Crump, K. S., and Shipp, A. M. (1988, December). Correlations between carcinogenic potency of chemicals in animals and humans. *Risk Analysis*, pp. 531–44.

Altman, Laurence K. (1990, 21 May). New method of analyzing health data stirs debate. *New York Times*, p. 31.

Altman, Laurence K. (1990, 29 May). The evidence mounts against passive smoking. *New York Times*, p. B5.

Aman, A. C., Jr. (1988). Administrative law in a global era: Progress, deregulation change, and the rise of the administrative presidency. *Cornell Law Review* 73:1101–1242.

American Iron & Steel Inst. v. OSHA, 577 F.2d 825 (1978).

American Textile Mfrs. Inst. v. Donovan, 452 U.S. 490–522 (1980).

Ames, B. (1989). Ranking possible carcinogenic hazards. *Science* 236:271–80.

Anderson, E. L. (1989). Scientific developments in risk assessment: Legal implications. *Columbia Journal of Environmental Law* 14:411–25.

Anderson v. Minneapolis, St.P. & S.St.M. R.R. Co., 146 Minn. 430, 179 N.W. 45 (1920).

ASARCO, Inc. v. OSHA, 746 F.2d 483 (9th Cir. 1984).

Baier, A. (1986). Poisoning the wells. In *Values at Risk*, ed. D. MacLean, Rowman & Allenheld; Totowa, N.J., pp. 49–75.

Bailar, J. C., Crouch, E. A. C., Shaikh, R., and Spiegelman, D. (1988). One-hit models of carcinogenesis: Conservative or not? *Risk Analysis* 8:485–97.

Barnard, R. C. (1986). An industry view of risk assessment. Paper presented at *Grappling with Risk Assessment: On the Frontier of Science and the Law*. Lake Tahoe, California.

Barnes, D. (1983). *Statistics as Proof*. Little, Brown, Boston.

Begging to defer: OSHA and the problem of interpretive authority (1989). *Minnesota Law Review* 73:1336. (note)

Bertrum v. Wenning, 385 S.W.2d 803 (Mo. App. 1965).

Black, B. (1987). Evolving legal standards for the admissibility of scientific evidence. *Science* 239:1510–12.

Black, B. (1988). A unified theory of scientific evidence. *Fordham Law Review* 56:595–692.

Black, B. & Lillienfeld, D. E. (1984). Epidemiologic proof in toxic tort litigation. *Fordham Law Review* 52:732–85.

Black's Law Dictionary, rev. 4th ed. (1968). West Publishing Company, St. Paul, Minn.

Borel v. Fiberboard Paper Products Corp., 493 F.2d 1076 (1973).

Brandt, R. (1979). *A Theory of the Good and the Right*. Clarendon Press, Oxford.

Bridbord, K., and French, T. (1978, 15 September). *Estimates of the fraction of cancer related to occupational exposures*. National Cancer Institute, National Institute of Environmental Health Sciences, National Institute of Occupational Safety and Health.

Broad, C. D. (1916). On the function of false hypotheses in ethics. *International Journal of Ethics* (now known as *Ethics*) 26:377 97.

Brodeur, P. (1985). *Outrageous Misconduct*. Pantheon, New York.

Broome, J. (1987). Utilitarianism and expected utility. *Journal of Philosophy* 84:405–22.

Brown, R. (1989) Issues in Federal Carcinogen Risk Assessment: The Food & Drug Administration. Paper presented at *Pesticides* and *Other Toxics: Assessing Their Risks*. University of California, Riverside.

Buchanan, A. (1984). The right to a decent minimum of health care. *Philosophy and Public Affairs* 13(1):36–56.

Building and Constr. Trades Dept., AFL-CIO v. Brock, 838 F.2d 1258 (D.C. Cir. 1988).

Bush, G. (1990). *Regulatory Program of the United States Government*. GPO, Washington, D.C.

Calabresi, G., and Hirschoff, J. (1972). Toward a test for strict liability in torts. *Yale Law Journal* 81:1054–85.

California Environmental Protection Agency, Health Hazard Assessment Division (1991, 26 April). Expedited method for estimating cancer potency: Use of tabulated TD_{50} values of Gold et al. Report presented to the Proposition 65 Science Advisory Panel.

California Environmental Protection Agency, Office of Environmental Health Hazard Assessment, Reproductive and Cancer Hazard Assessment Section (April 1992), "Expedited Cancer Potency Values and Proposed Regulatory Levels for Certain Proposition 65 Carcinogens."

California Health and Safety Code, §§ 25249.5 et seq., 7401–7671q.

Chevron U.S.A. Inc. v. Natural Resources Defense Council, Inc., 467 U.S. 837 (1983).

Clean Air Act (CAA), 1970 amended 1974, 1977, 1991. 42 U.S.C. §§ 7401–7671q.

Clean Water Act (CWA), formerly the Federal Water Control Act, 1972, amended 1977, 1978, 1987. 33 U.S.C. §§ 1251–1387.

Cohen, S. (1987). Knowledge, context and social standards. *Synthese* 73:3–20.

Coleman, J. L. (1976). Strict liability in torts. *William and Mary* Law Review 18:259.

Comprehensive Environmental Response, Compensation and Liability Act (CERCLA or "Superfund"), 1980, amended 1986. 42 U.S.C. §§ 9601–75.

Conn, R. (1980, February 3–10). "A Time Bomb Ticks Away In A Worker's Lungs." *The Charlotte* (N.C.) *Observer*.

Cornfield, J., Rai, K., and Van Ryzin, J. (1980). Procedures for assessing risk at low levels of exposure. *Archives of Toxicology* (Suppl. 3): 295–303.

Cothern, R. C., Coniglio, W. L., and Marcus, W. L. (1986). Estimating risks to health. *Environmental Science and Technology* 20(2)111–16.

Cox, L. A. and Dwyer, J. P. (1989). Acceptable cancer risks: Probabilities and beyond. Journal of the Air Pollution Control Association 39:1046, 1051.

Cranor, C. (1983). Epidemiology and procedural protections for workplace health in the aftermath of the *benzene* case. *Industrial Relations Law Journal* 5:372–400.

Cranor, C. (1985). Collective and individual duties to protect the environment. *Journal of Applied Philosophy* 2:243–59.

Cranor, C. (1985). Joint causation, torts and regulatory law in workplace health protections. *The International Journal of Applied Philosophy* 2:59–84.

Cranor, C. (1985). The justice of workplace health protections. *Logos* 6:131–48.

Cranor, C. (1986). Political philosophy, morality and the law. *Yale Law Journal* 95:1066–86.

Cranor, C. (1988). Some public policy problems with the science of carcinogen risk assessment. *Philosophy of Science* 2:467–88.

Cranor, C. (1990). Some moral issues in risk assessment. *Ethics* 101:123–43.

Cranor, C. (1991). Comparative risk judgements. *Protecting Drinking Water at the Source,*

J. J. De Vries University of California Water Resources Center, Davis, Calif. pp. 151–64.

Cranor, C., Byus, C., Chang, A., Gokhale, D. V., and Olsen, L. (1989). Final report to the University of California Toxic Substances Research and Teaching Program on the University of California, Riverside, Carcinogen Risk Assessment Project.

Cranor, C., and Nutting, K. (1990). Scientific and legal standards of statistical evidence in toxic tort and discrimination law. *Law and Philosophy* 9(2):115–56.

Cranor, C., Zeise L., Pease, W., Weiss, L., Hennig, M., and Hoover, S. (forthcoming in *Risk Analysis*). Improving the regulation of carcinogens: A proposal to expedite the estimation of cancer potency.

Daniels, N. (1981). Health care needs and distributive justice. *Philosophy and Public Affairs* 10(2):146–79.

Daniels, N. (1985). *Just Health Care*. Cambridge University Press, New York.

Delgado, R. (1982). Beyond Sindell: Relaxation of cause-in-fact rules for indeterminate plaintiffs. *California Law Review* 70:881–908.

Dworkin, R. (1986). *Law's Empire*. Belknap Press, Cambridge, Mass.

Dwyer, J. P. (1988). Overregulation. *Ecology Law Quarterly* 15:715–31.

Dwyer, J. P. (1990). The limits of environmental risk assessment. *Journal of Energy Engineering.* 116(3):231–2.

Eads, C., and Reuter, P. (1983). *Designing Safer Products: Corporate Responses to Product Liability Law and Regulation*. Rand Corp., Santa Monica, Calif.

Elam v. Alcolac, Inc., 765 S.W.2d 42 (Mo. App. 1988).

Elgin, C. (1988). The epistemic efficacy of stupidity. *Synthese* 72:297–311.

Environmental Health Letter (1989) 28(13):101. "Asbestos Ban Cleared. EPA Official Criticizes OMB for Costly Delays."

Epstein, S. S., and Swartz, J. B. (1981). Fallacies of lifestyle theories. *Nature* 289:127–30.

Estep, S. D. (1960). Radiation injuries and statistics: The need for a new approach to injury litigation. *Michigan Law Review* 59:259.

Evans v. S.J. Groves & Sons Co., 315 F.2d 335 (2d Cir. 1963).

Eysenck, H. J. (1991). Were we really wrong? *American Journal of Public Health.* 133(5):429–32.

Farland, W. (1990). Issues in federal carcinogen risk assessment: Environmental Protection Agency. In *Proceedings: Pesticides and Other Toxics: Assessing Their Risks,* ed. J. L. White. University of California, Riverside, College of Natural and Agricultural Sciences, pp. 127–36.

Feinberg, J. (1970). The expressive function of punishment. In *Doing and Deserving,* ed. J. Feinberg. Princeton University Press, Princeton, N.J., pp. 95–118.

Feinberg, J. (1970). Sua culpa. In *Doing and Deserving,* ed. J. Feinberg. Princeton University Press, Princeton, N.J., pp.187–221.

Feinstein, A. R. (1977). *Clinical Biostatistics,* Saint Louis: C. V. Mosby.

Feinstein, A. R. (1988). Scientific standards in epidemiologic studies of the menace of daily life. *Science* 242:1257–63.

Ferebee v. Chevron Chemical Co., 736 F.2d 1529 (D.C. Cir. 1984).

Finkel, A. M. (1989). Is risk assessment too conservative: Revising the revisionists. *Columbia Journal of Environmental Law* 14:427–67.

Finkel, A. M. (1990). *Confronting uncertainty in risk management: A report.* Center for Risk Management, Resources for the Future. Washington, D.C.

Fisher, A., Chesnut, L., and Violette, D. (1989). *Journal of Policy Analysis Management* 8:88.

Fleiss, J. L. (1986). Significance tests have a role in epidemiologic research: Reactions to A. M. Walker. *American Journal of Public Health* 76:559–60.

Food, Drug, and Cosmetic Act (FDCA), 1906, 1938, amended 1958, 1960, 1962, 1968. 21 U.S.C. §§ 301–94, 409.

Franklin, M. (1979). *Cases and Materials on Tort Law and Alternatives*. The Foundation Press, Mineola, N.Y.

Freedman, D., and Zeisel, H. (1988). From mouse to man: The quantitative assessment of cancer risks. *Statistical Science* 3:3–56.

Frye v. United States, 293 F.2d 1013 (D.C. Cir. 1923).

Furst, A. (1990). Yes, but is it a human carcinogen? *Journal of the American College of Toxicology* 9:1–18.

Gaines, S. E. (1988, 30 November). A taxonomy of toxic torts. *Toxics Law Reporter,* p. 826–31.

Giere, R. (1979). *Understanding Scientific Reasoning*. New York: Holt, Rinehart and Winston.

Gillette, C. P., and Krier, J. E. (1990). Risk, courts and agencies. *University of Pennsylvania Law Review* 138:1077–1109.

Gold, L., de Veciana, M., Backman, G., Magaw, R., Lopipero, P., Smith, M., Blumental, M., Levinson, R., Bernstein, L., and Ames, B. (1986). Chronological supplement to the Carcinogenic Potency Database: Standardized results of animal bioassays published through December 1982. *Environmental Health Perspectives* 67:161–200.

Gold, L., Sawyer, C. B., Magaw, R., Backman, G., de Veciana, M., Levinson, R., Hooper, N., Havender, W., Bernstein, L., Peto, R., Pike, M., and Ames, B. (1984). A carcinogenic potency database of the standardized results of animal bioassays. *Environmental Health Perspectives* 58:9–319.

Gold, L., Slone, T., Backman, G., Eisenberg, S., Da Costa, M., Wong, M., Manley, N., Rohrback, L., and Ames, B. (1990). Third chronological supplement to the Carcinogenic Potency Database: Standardized results of animal bioassays published through December 1986 and by the National Toxicology Program Through June 1987. *Environmental Health Perspectives* 84:215–85.

Gold, L., Slone, T., Backman, G., Magaw, R., Da Costa, M., and Ames, B. (1987). Second chronological supplement to the Carcinogenic Potency Database: Standardized results of animal bioassays published through December 1984 and by the National Toxicology Program through May 1986. *Environmental Health Perspectives* 74:237–329.

Goodman, S. N., and Royall, R. (1988). Evidence and scientific research (commentary). *American Journal of Public Health* 78:1568–74.

Greenland, S. (1991). Invited Commentary: Science versus Public Health actions: Those who were wrong are still wrong. *American Journal of Public Health* 133(5):435–36.

Greenland, S. (1990). Re: those who were wrong (letter). *American Journal of Epidemiology* 132(8):585–86.

Grinnell v. Pfizer Co., 274 Cal. App.2d 424, 70 Cal. Rptr. 375 (1969).

Grossman, E. (1990). Issues in federal carcinogen risk assessment: Occupational Safety and Health Administration. In *Proceedings: Pesticide Use in a Technological Society and Pesticides and Other Toxics: Assessing Their Risks,* ed. Janet L. White. University of California, Riverside, College of Natural and Agricultural Sciences, pp. 120–22.

Gulf S. Insul. v. Consumer Product Safety Commission, 701 F.2d 1137 (5th Cir. 1983).

Hall K. L. and Sibergeld E. K. (1983). Reappraising epidemiology: A response to Mr. Dore. *Harvard Environmental Law Review* 7(2) 441–48.

Hall, J. V., Winer, A. M., Kleinman, M. T. Lurmann, F. T., Brojus, V., Caloure, S. D. (1992). Valuing the health benefits of clean air. *Science* 255:812-17.

Harvard University School of Public Health. Center for Risk Analysis. (1991, June). *OMB v.*

the Agencies: The Future of Cancer Risk Assessment (conference). Cambridge, Massachusetts.

Haseman, J. K., Huff, J., Zeiger, E., McConnell, E. (1987). Comparative results of 327 chemical carcinogenicity studies. *Environmental Health Perspectives* 74:229–35.

Hattis, D. (1988). Seminar presentation at the University of California, Riverside, Carcinogen Risk Assessment Project, January 1988.

Hattis, D., and Smith, J. A., Jr. (1986). What's wrong with risk assessment? In *Quantitative Risk Assessment: Biomedical Ethics Reviews,* ed. J. M. Humber and R. F. Almeder. Humana Press, Clifton, N.J., pp. 57–106.

Holmes, L. B. (1986). Letter. *Journal of the American Medical Association* 256(22):3096.

Horowitz, R. I., and Feinstein, A. R. (1979). Methodologic standards and contradictory results in case-control research. *American Journal of Medicine* 66:556–63.

Huber, P. (1985). Safety and the second best: The hazards of public risk management in the courts. *Columbia Law Review* 85:277.

Huber, P. (1988). *Liability: The Legal Revolution and Its Consequences.* Basic Books, New York.

Hull, D. L. (1988). *Science as Process.* University of Chicago Press, Chicagao.

Humber, J. M, and Ameder, R. F., eds. (1988). *Quantitative Risk Assessment: Biomedical Ethics Reviews.* Humana Press, Clifton N.J.

Industrial Union Dept., AFL-CIO v. American Petroleum Inst., 448 U.S. 607 (1980).

Industrial Union Dept., AFL-CIO v. Hodgson, 499 F.2d 467 (1974).

Industrial Union Dept., AFL-CIO v. Marshall, 199 U.S. App. D.C. 54, 617 F.2d 636 (1979).

In re "Agent Orange" product liability litigation, 611 F.Supp. 1223, (C.D. N.Y. 1985).

In re Paoli R.R. Yard PCB litigation, 916 F.2d 829 (3d Cir. 1990).

James, F., and Hazard, G. (1977). *Civil Procedure,* 2d ed. Little, Brown, Boston.

Job-related murder convictions of three executives are overturned. (1990, January 20). *New York Times,* p. 10.

Johnson v. U.S., 597 F. Supp. 374 (D. Kan. 1984).

Johnson, B. B., and Covello, V. T., eds. (1987). *The Social and Cultural Construction of Risk.* D. Reidel, Bingham, Mass.

Johnson, W. M., and Parnes, W. D. (1979). Beta-naphthalamine and benzidine: Identification of groups at high risk of cancer. In *Annals of the New York Academy of Science: Public Control of Environmental Health Hazards,* ed. E. Cuyler Hammond & Irving J. Selikoff. New York Academy of Sciences, New York, pp. 277–84.

Kahn, S. (1989). Economic estimates of the value of human life, in *Ethics & Risk Management in Engineering,* ed. A. Flores. University Press of America, New York, pp. 57–72.

Karstadt, M. (1986). Quantitative risk assessment: Qualms and questions. Paper presented at *Grappling with Risk Assessment: On the Frontier of Science and the Law.* Lake Tahoe, California.

Kavka, G. (1978). The numbers should count. *Philosophical Studies* 36:285–94.

Keeton, W. P., Dobbs, D. B., Keeton, R. E., and Owen D. G., (1984). *Prosser and Keeton on Torts,* 5th ed. West Publishing Co. St. Paul, Minn.

Kelsey, J. L., Thompson, W. D., Evans, A. S. (1986). *Methods in Observational Epidemiology.* Oxford University Press, New York, pp. 128–30.

Koltat, G. (1990, 31 August). Scientists question methods used in cancer tests. *New York Times,* p. 1.

Kogevinas, M., and Boffetta, P. (1991). Letter. *British Journal of Industrial Medicine* 48:575–76.

Krewski, D., Szyszkowicz, M., and Rosenkranz, H. (1990). Quantitative factors in chemical carcinogenesis: Variations in carcinogenic potency. *Regulatory Toxicology and Pharmacology* 12:13–29.

Kronman, A. (1979). Mistake and impossibility: Arranging a marriage between two different partners 7. *Northwestern University Law Review* 74:166–201.

Latin, H. (1983). The feasibility of occupational health standards: An essay of legal decision making under uncertainty. *Northwestern University Law Review* 78:583–631.

Latin, H. (1988). Good science, bad regulation and toxic risk assessment. *Yale Law Journal on Regulation* 5:89–142.

Lave, L. Omenn, G., Heffernan, K., and Dranoff, G. (1983). A model for selecting short-term tests for carcinogenicity. *Journal of the American College of Toxicology* 2:125–30.

Lave, L., Ennever, F., Rosenkranz, H., and Omenn, G. (1988). Information value of the rodent bioassay. *Nature* 336:631–33.

Lave, L., and Omenn, G. (1986). Cost effectiveness of short-term tests for carcinogenicity. *Nature* 324:29–34.

Lave, L., and Omenn, G. (1988). Screening toxic chemicals: How accurate must tests be? *Journal of the American College of Toxicology* 7:565–74

Lawless, E. W. (1977). *Technology and Social Shock.* Rutgers University Press, New Brunswick, N.J.

Lee, D. Y., & Chang, A. C. (1992). Evaluating transport of organic chemicals in soil resulting from underground fuel tank leaks. *46 Purdue Industrial Waste Conference Proceedings.* Lewis Publishers, pp. 131–40.

Lehman, E. L. (1959). *Testing Statistical Hypotheses.* Wiley, New York.

Levy, E., and Copp, D. (1983). Risk and Responsibility: Ethical issues in decisionmaking. *Contemporary Moral Issues,* ed. W. Cragg. Toronto: McGraw-Hill, Ryerson.

Localio, A. R., Lawthers, A. G., Brennan, T., Laird, N. M., Hebert, L. E., Peterson, L. M., Newhouse, J. P., Weiler, P. C., Haitt, H. H. (1991). Relation between malpractice claims and adverse events due to negligence. *New England Journal of Medicine* 325:245–51.

Lowe, M. F. (1989). Risk assessment and the credibility of federal regulatory policy: An FDA perspective. *Regulatory Toxicology and Pharmacology* 9:131–40.

Lynch v. Merrill-National Laboratories, 830 F.2d 1190 (1st Cir. 1987).

Lynn F. M. (1987). OSHA's carcinogen standard: Round one on risk assessment models and assumptions. In *The Social and Cultural Construction of Risk,* ed. B. B. Johnson & V. T. Covello. D. Reidel, Bingham, Mass., pp. 345–58.

Lyons, D. (1977). Human rights and general welfare. *Philosophy and Public Affairs* 6:113–29.

MacLean, D. (1986). Social values and the distribution of risk. In *Values at Risk,* ed. D. MacLean. Rowman & Allenheld, Totowa, N.J.

Mashaw, J. (1985). Prodelegation: Why administrators should make political decisions. *Journal of Law, Economics and Organization* 1:82.

Mashaw, J. L., and Merrill, R. A. (1985). *Administrative Law: The American Public Law System,* 2d ed. West Publishing Co., St. Paul, Minn., pp. 416–23.

Maugh, T. H. (1990, 31 August). Carcinogen test process challenged. *Los Angeles Times,* p. 1.

Mausner, J. S., and Bahn, A. K. (1974). *Epidemiology: An Introductory Text.* W. B. Saunders, Philadelphia.

Mayes, L. C., Horowitz, R. I., and Feinstein, A. R. (1988). A collection of 56 topics with contradictory results in case-control research. *International Journal of Epidemiology* 17:680–85.

Mayo, D. (1988). Toward a more objective understanding of the evidence of carcinogenic risk. *Philosophy of Science Association* 2:489–503.

McAllister v. Workmen's Compensation Appeals Board, 445 P.2d. 313 (Calif. Sup. Ct. 1968).

Cleary, E. W., Ball, V. C., Barnhart, R. C., Broun, K. S., Dix, G. E., Gelhorn, E., Meisenholder, R., Roberts, E. F., and Strong, J. C. (1984). *McCormick on Evidence.* West Publishing Co., St. Paul, Minn.

McGarity, T. (1987). Regulatory analysis and regularity reform. *Texas Law Review,* 1243–1333.

Merrill, R. A. (1979). Regulating carcinogens in food: A legislator's guide to the food safety provisions of the Federal Food, Drug and Cosmetic Act. *Michigan Law Review* 72:172–250.

Mills, J. L., and Alexander, D. (1986). Letter. *New England Journal of Medicine* 315:1234.

Motor Vehicle Mfrs. Assn. v. State Farm Mut., 463 U.S. 29 (1982).

Murphy, J. G., and Coleman, J. L. (1984). *The Philosophy of Law: An Introduction to Jurisprudence.* Rowman and Allenheld, Totowa, N.J.

National Research Council. (1983). *Risk assessment in the federal government: Managing the process.* GPO, Washington, D.C.

National Cancer Institute. Demographic Analysis Section. Division of Cancer Cause and Prevention. (1981). *Surveillance, epidemiology and end results: Incidence and mortality data 1973–1977* (monograph 57).

National Research Council. (1984). *Toxicity Testing: Strategies to Determine Needs and Priorities.* National Academy Press, Washington, D.C.

National Research Council. (1987). *Pharmacokinetics in Risk Assessment, Drinking Water and Health.* National Academy Press, Washington, D.C.

National Toxicology Program. Division of Toxicology Research and Testing. (1991, 7 July). *Chemical Status Report.*

Needleman, H., and Billings, D. (forthcoming) Type II fallacies in the study of childhood exposure to lead at low dose: A critical and quantitative review.

Nelson, J. (1983, 8 May). Critics cry foul on health, safety. *Los Angeles Times.* p. 1.

New Jersey Supreme Court applies broader test for admitting expert testimony in toxic case. (1991, 27 August). *Environmental Health Letter* 30:171–80.

NIOSH Criteria Document. (1972). Occupational exposure to asbestos. Ch. 3, 7. U.S. Dept. of Health and Human Services, Public Health Service DHHS publication.

Occupational Safety Health Act (OSHA), 1970. 29 U.S.C. §§ 651–78.

O'Leary, R. (1989). The impact of federal court decisions on the policies and administration of the U.S. Environmental Protection Agency. *Administrative Law Review* 41:549–78.

Olson, L. J. (1990). The search for a safe environment: The economics of screening and regulating environmental hazards. *Journal of Environmental Economics and Management* 19:1–18.

Packer, H. (1968). *The Limits of the Criminal Sanction.* Stanford University Press, Palo Alto, Calif.

Page, T. (1978). A generic view of toxic chemicals and similar risks. *Ecology Law Quarterly* 7(2) pp. 207–44.

Page, T. (1981). A framework for unreasonable risk in the Toxic Substances Control Act (TSCA). In *Annals of the New York Academy of Science: Management of Assessed Risk for Carcinogens,* ed. W. J. Nicholson. New York Academy of Science, New York, pp. 145–66.

Page, T. (1983). On the meaning of the preponderance of the evidence test in judicial regulation of chemical hazards. *Law and Contemporary Problems* 46:267–83.

Page, T. (forthcoming) The significance of P-values.

Peto, R., Pike, M., Bernstein, L., Gold, L., & Ames, B. (1984). The TD^{50}: A proposed general convention for the numerical description of the carcinogenic potency of chemicals in chronic-exposure animal experiments. *Environmental Health Perspectives* 58:1–8.

Pierce, R. J., Jr. (1980). Encouraging safety: The limit of the tort law and government regulation. *Vanderbilt Law Review* 33:1281–1331.

Pierce, R. J., Jr. (1988). Two problems in administrative law: Political polarity on the District of Columbia circuit and judicial deterrence of agency rulemaking. *Duke Law Journal*. 1988:300–28.

Plater, Z. J. B., and Norine, W. L. (1989). Through the looking glass of eminent domain: Exploring the "arbitrary and capricious" test and substantive rationality review of governmental decisions. *Boston University Journal of Environmental Affairs* 16:661–717.

Poole, C. (1987). Beyond the confidence interval. *American Journal of Public Health* 77:195–99.

The President's Council on Competitiveness (1991). GPO, Washington, D.C.

Public Citizen v. Young, 831 F.2d 1108 (D.C. Cir. 1987).

Public Citizen Health Research Group v. Tyson, 796 F.2d 1479 (D.C. Cir. 1986).

Quine, W. V. O. (1960). *Word and Object.* MIT Press, Cambridge, Mass.

Rall, D. P. (1979). The role of laboratory animal studies in estimating carcinogenic risks for man. In *Carcinogenic Risks: Strategies for Intervention.* International Agency for Research on Cancer, Lyon, France.

Rall, D. P. (1980). Testimony. *Federal Register* 45:5061.

Rawls, J. (1971). *A Theory of Justice.* Harvard University Press, Cambridge, Mass.

Reagan, R. (1981, 17 February). Federal regulation. Executive order 12,291. *Federal Register* 46:13,193.

Reagan, R. (1985, 4 January). Regulatory planning process. Executive order 12,498. *Federal Register* 50:1036.

Rescher, N. (1966). *Distributive Justice.* Bobbs-Merrill, New York.

Rescher, N. (1983). *Risk: A Philosophical Introduction to the Theory of Risk Evaluation and Management.* University Press of America, Washington, D.C.

Resource Conservation and Recovery Act (RCRA), 1976. 42 U.S.C. §§ 6901–92.

Ricci, P. F., Cox, L. A., and Dwyer, J. P. (1989). Acceptable cancer risks: Probabilities and beyond. *Journal of the Air Pollution Control Association* 39:1046.

Richardson' by Richardson v. Richardson-Merrill, Inc., 649 F. Supp. 799 (D.D.C. 1986), *affd.* at 857 F.2d 823 (D.C. Cir. 1989) *cert. denied* 110 S. Ct. 218 (1989).

Rimm, A. (1980). *Basic Biostatistics in Medicine and Epidemiology.* New York: Appleton Century Crofts.

Rodgers, W. H., Jr. (1977). *Handbook on Environmental Law.* West Publishing Co., St. Paul, Minn.

Rosenberg, D. (1984). The causal connection in mass exposure cases: A "public law" vision of the tort system. *Harvard Law Review* 97:849–929.

Rosenberg, D. (1986). The dusting of America, A story of asbestos—carnage, cover-up and litigation. *Harvard Law Review* 99:1693–1706.

Ross, W. D. (1970). The right and the good. In *Readings in Ethical Theory,* 2d ed. Sellars, W. and Hospers, J. Appleton-Century-Crofts, New York.

Rothman, K. J. (1986). *Modern Epidemiology.* Little, Brown, Boston.

Rubanick v. Witco Chemical Corp. & Monsanto Co. (N.J. Sup. Ct., 1991).

Rubanick v. Witco Chemical Corp. 576 A.2d 4 (N.J. Sup. A.D., 1990).

Safe Drinking Water Act (SDWA), 1976, amended 1986. 42 U.S.C. §§ 300f–300J–26.

Sagoff, M. (1988). *The Economy of the Earth.* Cambridge University Press, New York.

Sanders, J. T. (1988). The numbers should sometimes count. *Philosophy and Public Affairs* 17(1):3–14.

Scanlon, T. M. (1973). Rawls' theory of justice. *Pennsylvania Law Review* 121:1020–69.

Scanlon, T. M. (1975). Preference and urgency. *Journal of Philosophy* 77:655–69.

Scanlon, T. M. (1982). Contractualism and utilitarianism. In *Utilitarianism and Beyond,* ed. A. Sen and B. Williams. Cambridge Univeersity Press, Cambridge, pp. 101–28.

Scheffler, S. (1982). *The Rejection of Consequentialism: A Philosophical Investigation of the Considerations Underlying Rival Moral Conceptions.* Oxford University Press, New York.

Schlesselman, J. J. (1974). Sample size requirements in cohort and care-control studies of disease. *American Journal of Epidemiology* 99:381–84.

Schmahl, D., Preussman, R., and Berger, M. R. (1981). Causes of cancer—an alternative view to Doll and Peto. *Klinische Wochenschrift* 67:1169–73.

Schneider, K. (1990, 3 May). Scientist who managed to "shock the world" on atomic workers health. *New York Times.* p. A20.

Schottenfeld D. and Haas J. F. (1979). Carcinogens in the workplace. *CA–Cancer Journal for Clinicians* 29 144–59.

Schwing, R. C. (1989, January). SRA: An antidote to symbols of insanity. *RISK Newsletter,* p. 2.

Scott, J. (1990, 1 August). Genes may increase the risk of lung cancer. *Los Angeles Times,* p. A11.

Scott, J. (1990, 31 August). Job-related illness called "America's invisible killer." *Los Angeles Times,* p. A4.

Shrader-Frechette, K. (1985). *Science Policy, Ethics and Economic Methodology.* D. Reidel, Boston.

Seilkin, R. L. (1986). Benzene and leukemia: Opposed to the shortcomings in the current regulatory process. Paper presented at *Grappling with Risk Assessment: On the Frontier of Science and the Law.* Lake Tahoe, California.

Selewski v. Williams, 72 Mich. App. 742, 250 N.W.2d 529 (1977).

Selikoff, I. J., Jacob Chung, and E. Cuyler Hammond. (1964). Asbestos Exposure and Neoplasia. *Journal of American Medical Association* 188:22–27.

Sen, A. and Williams, B., eds. (1982). *Utilitarianism and Beyond.* Cambridge University Press, New York.

Shavell, S. (1984). Liability for harm versus regulation of safety. *Journal of Legal Studies* 13:357–71.

Shavell, S. (1987). *Economic Analysis of Accident Law.* Harvard University Press, Cambridge, Mass.

Silkwood v. Kerr-McGee Corp., 485 F.Supp. 566 (W.D. Okla. 1979).

Sindell v. Abbot Laboratories et al., 26 Cal.3d 588, 607 P.2d 924 (1980).

Slovic, P. C. (1987). Perception of risk. *Science* 236:280–85.

Smart, J. J. C., and Williams, B. (1973). *Utilitarianism: For and Against.* Cambridge University Press, Cambridge.

Smith v. Lockheed Propulsion Co., 247 Cal. App.2d 774 (56 Cal. Rptr. 128, 1967).

Society of Plastics Industry, Inc. v. OSHA., 509 F.2d 1301 (1975).

Steinhauser v. Hertz Corp., 421 F.2d 1169 (2d Cir. 1970).

Sterling v. Velsicol Chemical Corp., 647 F. Supp. 303 (W.D. Tenn. 1986).

Sterling v. Velsicol Chemical Corp., 855 F.2d 718 (6th Cir. 1988).

Stevens, W., Till, J. E., Thomas, D. C., Lyon, J. L., and Kerber, R. A. (forthcoming). A historical reconstruction and epidemiologic study of leukemia in Utah residents exposed to fallout.

Stewart, R. B. and Krier, J. E. (1978). *Environmental Law and Policy* Bobbs-Merrill, New York.

Strauss P. l. (1987). One hundred fifty cases per year: Some implications of the supreme court's limited resources for judicial review of agency action. *Columbia Law Review* 87:1093–1136.

Sugarman, S. D. (1990). The need to reform personal injury law leaving scientific disputes to scientists. *Science* 248:823–27.

Summers v. Tice, 33 Cal.2d 80, 199 F.2d 1 (1948).

Synthetic Organic Mfrs. Ass'n. v. Brennan, 503 F.2d 1155 (3d Cir. 1974).

Taurek, J. (1977) Should the numbers count? *Philosophy and Public Affairs.* 6(4):293–316.

Texas Independent Ginners Ass'n. v. Marshall, 630 F.2d 398 (5th Cir. 1980).

The President's Council on Competitiveness (1991). GPO, Washington, D.C.

Tomatis, L. (1988). Environmental cancer risk factors: A review. *Acta Oncologica* 27:465–72

Tomatis, L. Editor, co-eds. Aitio, A., Day, N. E., Heseltine, E., Kaidor, J., Miller, A. B., Parkin, D. M., Riboli, E. (1990). *Cancer: Causes, Occurrence and Control.* International Agency for Research on Cancer, Lyon, France. New York: Distributed in the USA by Oxford University Press.

Thompson, W. D. (1987). Statistical criteria in the interpretation of epidemiologic data. *American Journal of Public Health* 77:191–94.

Toxic Substances Control Act (TSCA), 1976. 15 U.S.C. §§ 2601–71.

Underwood, B. (1977). The thumb on the scales of justice: Burdens of persuasion in criminal cases. *Yale Law Journal* 86:1299.

U. S. Congress. (1981). Office of Technology Assessment. *Assessment of Technologies for Determining Cancer Risks from the Environment.* GPO, Washington, D.C.

U. S. Congress. Office of Technology Assessment. (1982). *Cancer Risk: Assessing and Reducing the Dangers in Our Society.* Westview Press, Boulder, Colo.

U. S. Congress. Office of Technology Assessment. (1987). *Identifying and Regulating Carcinogens.* GPO, Washington D.C.

U. S. Department of Labor. OSHA. (1980). Identification, classification, and regulation of potential occupational carcinogens. *Federal Register* 45:5002–5294.

U. S. Interagency Regulatory Liaison Group. (1979). Scientific bases for identification of potential carcinogens and estimation of risks. *Journal of the National Cancer Institute* 63(1):241–60.

U. S. Interagency Staff Group on Carcinogens. (1986). Chemical carcinogens: A review of the science and its associated principles. *Environmental Health Perspectives* 67:201–82.

United Steelworkers, AFL-CIO v. Brennan, 530 F.2d 109 (3d Cir. 1975).

United Steelworkers v. Marshall, 647 F.2d 1189 (D.C. Cir. 1980).

Universal Camera Corp. v. National Labor Relations Board, 340 U.S. 474 (1951).

Van Ryzin, J. (1982). Current topics in biostatistics and epidemiology (discussion). *Biometrics* 28:130–38.

Van Ryzin, J., and Rai, K. (1980). The use of quantal response states to make predictions. In *The Scientific Basis of Risk Assessment,* ed. H. Witschi, New York, N.Y., Elsevier/North Holland Biomedical Press, pp. 273–90.

Walker, A. M. (1986). Reporting the results of epidemiologic studies. *American Journal of Public Health* 76:556–58.

Walter, S. D. (1977). Determination of significant relevant risks and optimal sampling procedures in prospective and retrospective comparative studies of various sizes. *American Journal of Epidemiology* 105:387–97.

Watkins, R. N. (1986). Letter. *Journal of the American Medical Association* 256:22.

Wells v. Ortho Pharmaceutical Corp., 788 F.2d 749 (1986).

Wong, O. (1990). A cohert mortality study and a case central study of workers potentially exposed to styrene in reinforced plastics and composite industry. *British Journal of Industrial Medicine* 47:753–62.

Roberts, L. (1989). Is risk assessment conservative? *Science* 243:1553.

Zeise, L. (1990). Issues in state risk assessment: California Department of Health Service. In *Proceedings: Pesticides and Other Toxics: Assessing Their Risks,* ed. Janet L. White. University of California, Riverside, College of Natural and Agricultural Sciences, pp. 135–44.

Zeise L., Pease, W., and Kelter, A. (1989, January). Risk assessment for carcinogens under California's Proposition 65. Presented at the Western Meetings of the American Association for the Advancement of Science, San Francisco, CA.

Zeise, L., Wilson, R., and Crouch, E. A. C. (1987). Dose–response relationships for carcinogens: A review. *Environmental Health Perspectives* 73:295–308.

Index

f = figure; t = table

Uncertainty (*Cont.*)
  in risk assessment, 20, 21, 23, 25, 28, 48,
    136, 146, 221
  scientific, institutional, and policy responses
    to, 28, 131–35, 151
  use of policy choices to address, 131–35,
    151, 157
Uncertainty ranges
  invite court intervention and misunderstand-
    ing, 122
  problem of understanding, 122
  use of, in numerical risk estimates, 121
Underregulation, 8. *See also* False negatives;
    Mistakes
  costs of, 8, 121, 124, 130, 142–46
  defined, 8
  and dose-response assessment, 14
  impact of evidentiary standards on, 11
  importance relative to overregulation, 126,
    129
  major, specified, 125
  minor, specified, 125
Underregulation and overregulation balance
  for each substance considered, 130
  for risk assessment process as whole, 130
U.S. Circuit Court of Appeals for District of
    Columbia, 56, 148, 150, 160–61
  interpretation of Occupational Safety and
    Health Act, 160–62
  upheld OSHA regulations on asbestos against
    industry challenge, 150
U.S. Court of Appeals for Second Circuit, 87
U.S. Court of Appeals for Fifth Circuit, 52, 122
U.S. Court of Appeals for Ninth Circuit, 149
U.S. Court of Appeals for Eleventh Circuit, 60
U.S. Environmental Protection Agency, 111,
    133
  cancer potency estimates of CDHS and EPA,
    222
  frequency distribution between EPA and Cali-
    fornia Department of Health Services po-
    tency values, 22*f*
  negotiated rulemaking, viii
  pesticide program, viii
  risk assessment practices, 113
U.S. Environmental Protection Agency, Carcin-
    ogen Assessment Group, 21, 27, 109
  risk assessments performed by, 22, 27

U.S. Supreme Court, 7, 111, 137–37, 149,
    162–63
Utilitarianism, 126, 134, 166, 176
  characterized, 163–64
  difficulty guaranteeing a certain distribution of
    resources, 167
  discussion of, 163–68
  does not take separateness of individual per-
    sons seriously, 166
  equations describing relationships between
    health protections, personal income, and
    employment opportunities, 164
  has greater difficulty assigning prominence to
    workplace and environmental health protec-
    tions than OSH Act or Daniels-Rawls view,
    175
  many may use for justifying use of toxic sub-
    stances, 177
  permits minor benefits to many to outweigh
    severe harms to few, 166
  preventive health protections have no special
    weight, 165
  problematic model of the good for human be-
    ings, 167
  some shortcomings of, 165
  weighing costs of false positives and false
    negatives, 168

Velsicol Chemical Corporation, 77

Weinstein, Judge Jack B., 56
Wells v. Ortho Pharmaceutical Corp., 60, 62
Workplace
  cancer from exposure in, 4
  concentrations of toxic substances in, 6
  health protections in, 158
  measurement of toxic substances in, 6
  as model for studying environmental health
    harms, 6
  moral and legal issues in, 7
Workplace and environmental health protections
  part of the background features of commu-
    nity life, 159, 171

Xylene, 20

Zeise, Lauren, 22*f*
Zeisel, H., 114